GAS-CHROMATOGRAPHIC ANALYSIS OF TRACE IMPURITIES

STUDIES IN SOVIET SCIENCE

GAS-CHROMATOGRAPHIC ANALYSIS OF TRACE IMPURITIES

V. G. Berezkin and V. S. Tatarinskii

A. V. Topchiev Institute for Petrochemical Synthesis
Academy of Sciences of the USSR
Moscow, USSR

Translated from Russian by
J. E. S. Bradley

Senior Lecturer in Physics
University of London
London, England

CONSULTANTS BUREAU • NEW YORK — LONDON • 1973

Viktor Grigor'evich Berezkin was born in 1931 in Moscow. In 1954 he was graduated from Moscow University. He presented his Ph.D. thesis in 1962 and in 1968 received his D.Sc. He is presently director of a laboratory at the Topchiev Institute of Petrochemical Synthesis, Academy of Sciences of the USSR.

Vladimir Sergeevich Tatarinskii was born in 1939 in Leningrad. After graduating from the Mendeleev Moscow Chemical Technology Institute, he worked in the Topchiev Institute of Petrochemical Synthesis on gas-chromatographic determination of trace impurities in organic compounds. He presented his Ph.D. thesis on this topic under the direction of V. G. Berezkin in 1969. He is presently a senior scientific associate at a research institute of the Ministry of the Chemical Industry of the USSR.

The original Russian text, published by Nauka Press in Moscow in 1970, has been corrected by the authors for the present edition. This translation is published under an agreement with Mezhdunarodnaya Kniga, the Soviet book export agency.

Газо-хроматографические методы анализа примесей

GAZO-KHROMATOGRAFICHESKIE METODY ANALIZA PRIMESEI

V. G. Berezkin and V. S. Tatarinskii

Library of Congress Catalog Card Number 72-94824

ISBN-13: 978-1-4684-1601-5   e-ISBN-13: 978-1-4684-1599-5
DOI: 10.1007/978-1-4684-1599-5

© 1973 Consultants Bureau, New York
Softcover reprint of the hardcover 1st edition 1973

A Division of Plenum Publishing Corporation
227 West 17th Street, New York, N. Y. 10011

United Kingdom edition published by Consultants Bureau, London
A Division of Plenum Publishing Company, Ltd.
Davis House (4th Floor), 8 Scrubs Lane, Harlesden, London NW10 6SE, England

Preface to the English Edition

Recent scientific and industrial developments have demonstrated the major effects (advantageous or otherwise) of trace components in compounds, which in commercial products may substantially affect the properties and the course of reactions, especially catalytic ones. Some of these problems (e.g., the purity of natural waters and air) have attracted the attention not only of specialists but also of society generally.

It would be impossible to do research on trace components and contaminants if we did not have reliable rapid and sensitive methods of determining substances present in mixtures in minute concentrations. Gas chromatography is an effective method of purity testing. This book deals with analysis for trace components, together with the basic methods of gas chromatography that are used for the purpose. We have tried to reflect here the principal papers in this area appearing in the foreign literature and in Russian.

We are indebted to Plenum Press for the opportunity of presenting this edition.

Moscow

September 1972

Preface

Gas-chromatographic analysis for impurities is presently a rapidly developing area in gas chromatography, because many practical topics give considerable importance to impurity analysis.

Progress in this analysis has been such for volatile compounds (and also for nonvolatile ones, when reactive gas chromatography is used) that one can now be confident of solving most practical problems in this way. Commercial ionization detectors of high sensitivity and selectivity often allow one to detect impurities down to levels of parts in 10^9.

There are good prospects for raising the sensitivity in gas chromatography by thermal methods that enrich the sample considerably in impurities (temperature programming, chromatothermography, etc.), by preliminary concentration of impurities (accumulation of impurities or removal of the principal component from large volumes of material), or by reactive gas chromatography.

Although there are many papers on the solution of particular problems, we consider that too little attention has been given in the literature to the specific problems of analysis for impurities. Also, there are no books that deal in detail with all aspects of gas-chromatographic determination of impurities.

We have written this as one on methods, not on narrow techniques; we present systematically the main results on the theory, on general methods of analysis, and on methods of impurity concentration. Detailed techniques are considered, as a rule, only to illustrate the methods.

We hope that the reader will find the book of interest and value.

Chapters III, IV, and V were written by the authors jointly; Chapters I, II, VI, VII (Section 1), and VIII were written by Berezkin, and Chapters VII (Section 2) and IX were written by Tatarinskii.

We are indebted to A. A. Zhukovitskii, A. F. Shlyakhov, I. A. Revel'skii, A. S. Bel'fer, Z. V. Verkhovskaya, S. A. Volkov, and V. I. Kalmanovskii for critical reading of various chapters in draft.

Contents

Introduction

The purity of a substance often affects many physical and chemical properties of a material (e.g., the electrical conductivity, luminescence, capacity for polymerization, radiation stability); often, components at the 10^{-2} to $10^{-9}\%$ level greatly affect the observed properties, so contaminated compounds cannot be used in industry or in scientific research. For instance, polymerization is widely used on the multiton scale in petrochemical and other industries, and only very pure compounds (monomers, solvents, etc.) can be used in such processes. Even small amounts of impurity affect not only the polymerization but also the properties of the polymers, especially thermal decomposition, particularly when this is of chain type. The maximum concentration should usually not exceed 10^{-2} to 10^{-4} % (in accordance with the reactivity), i.e., in a polymerization one can allow only from a few single molecules of impurity up to a few hundred per million of the monomer [2, 3].

The correct approach in any scientific research is to use pure compounds under conditions that rule out contamination. Major areas in modern analytical chemistry are therefore analysis of contaminants in pure compounds and determination of impurities in commercial products [4, 5]. Nogare and Juvet [6] define an impurity as a trace component if its concentration in the mixtures is equal to or less than $10^{-2}\%$.

Volatile impurities in monomers and other very pure compounds can be determined with advantage by gas chromatography, for the following reasons:

1. The method gives in a single analysis information about many contaminants, not merely a single one, particularly since it gives very sharp separations, which allows one to analyze for impurities that differ only slightly in properties (isomers, etc.);
2. Sensitive detectors allow one to detect impurities at very low concentrations, and accumulation can be used independently to reduce further the minimum detectable concentration;
3. The method is rapid, as the times needed for one analysis range from a few minutes to around an hour;
4. Standard apparatus can be used.

These advantages have led to general use of the method in the analysis of high-purity compounds [7, 8]. The literature gives many examples of analyses of particular substances, in particular monomers, drugs, metals, food products, air (and other gases), biological materials (blood, urine), etc.

Impurity analysis is now a major independent area in gas chromatography, which has special methods of analysis and of quantitative evaluation of the results, with general use of accumulation techniques. It is also an area with fresh potential sources of error.

We give especial attention to general methods and approaches for components present in trace amounts, as well as to the specific features and difficulties of gas-chromatographic analysis for trace components.

LITERATURE CITED

1. V. A. Kargin, Basic Problems in Polymer Chemistry [in Russian], Moscow, Izd. AN SSSR (1958).
2. S. E. Bresler and B. L. Erusalimskii, Physics and Chemistry of Macromolecules [in Russian], Moscow Leningrad, Nauka (1965).
3. L. S. Kofman and V. S. Vinogradov, Izv. AN SSSR, Ser. Khim., 975 (1965).
4. P. Auger, Current Trends in Scientific Research, UNESCO (1963).
5. Analysis Methods for High-Purity Substances [in Russian], edited by I. P. Alimarin, Moscow, Nauka (1965).
6. S. D. Nogare and R. S. Juvet, Gas–Liquid Chromatography [Russian translation], Leningrad, Nedra (1966).

7. Gas Chromatography: Literature Survey 1952-1960 [in Russian], Moscow, Izd. AN SSSR (1962).

8. Gas Chromatography: Literature Survey 1961-1966 [in Russian], Moscow, Nauka (1969).

Chapter I

Gas-Chromatographic Separation in Impurity Analysis

The amount of the main component exceeds that of the impurities by factors of 10^4-10^8 in such analyses, and these unusual conditions may appreciably affect the qualitative and quantitative results.

A typical chromatogram (Fig. 1) illustrates some features of impurity analysis. The main component gives an off-scale reading, and its width substantially exceeds the impurity widths. The main peak also has a pronounced tail. This makes it impossible to determine impurities whose retention time is in the off-scale region of the main peak, and there is also reduced accuracy for impurities whose peaks fall on the tail.

Overlap from the main peak is the most frequent cause of complication in gas-chromatographic analysis. Very often, only certain impurities have to be assayed, and this allows the following classification of the problems:

1. The retention times t_r for those impurities are very different from that of the base:

$$t_{rj} \gg t_{ba} \gg t_{ri},$$

where t_{ba} is the base retention time, t_{ri} is the retention time for the fast impurities, and t_{rj} is the same for the slow ones;

2. The time differences are not very pronounced:

$$t_{rj} > t_{ba} > t_{ri}.$$

5

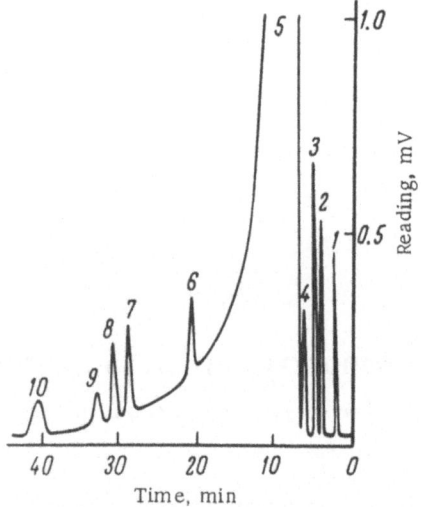

Fig. 1. Chromatogram of impurities in cyclohexane:
1) pentane, 2) hexane, 3) methylcyclopentane,
6) toluene, 7) octane, 8) ethylcyclohexane, 9) 2,6-
dimethylhept-1-ene, 10) nonane. 270 × 0.4 cm column, 10% squalene on celite 545, 75°C, 50 ml/min
of nitrogen, sample volume 2 μl.

3. The times are comparable:

$$t_{rj} \approx t_{ba} \approx t_{ri}.$$

The second and third cases are usually the most troublesome,
since the base peak then overlaps the other peaks.

It is now usually fairly straightforward to separate the components of a complex mixture when these are present in comparable amounts [1-3, 49, 50], so it is best to establish how far the con-
cepts and trends are applicable in trace-component analysis. In
particular, we are interested in the case where the impurities are
well separated from the base.

Two opposing processes occur as the carrier gas moves the
components along the sorbent layer: the peaks for adjacent components move apart (which improves the separation) and the peak
width increases (which has the opposite effect).

The separation between two compounds is governed by two characteristics: the sorbent selectivity, which can be evaluated as the ratio of the retention times t_2 and t_1,

$$\alpha = \frac{t_2}{t_1} \qquad (1)$$

and the column performance, which can be evaluated as the number N of theoretical plates,

$$N = 16\left(\frac{t}{\mu}\right)^2, \qquad (2)$$

where t is retention time and μ is the width of a peak at the bottom. The separation increases with N and α.

If two compounds are present in comparable concentrations, there are various ways of characterizing the separation quantitatively in terms of t and μ [2, 4-6]. A committee [6] recommended the following for this:

$$R = \frac{2(t_2 - t_1)}{\mu_2 + \mu_1} = \frac{2\Delta t}{\mu_2 + \mu_1}, \qquad (3)$$

where Δt is the separation between the two peaks and μ_1 and μ_2 are the base widths of the peaks. This R varies from 0 to ∞, and the peaks are completely separate for R = 1. The following quantity is commonly used in the Russian literature [2, 7] for the separation criterion:

$$K_1 = \frac{1}{2}R. \qquad (4)$$

This K_1 is determined by α and N [8]:

$$K_1 = \frac{\sqrt{2}\,(m-1)}{4\,(m+1)}\sqrt{N} = \frac{\sqrt{2}}{4}K_c\sqrt{N}, \qquad (5)$$

where $K_c = (m-1)/(m+1)$ and $m = \Gamma_1/\Gamma_2$, and Γ is the partition coefficient, which characterizes the sorption of a compound by unit volume of the sorbent. We can take $m \approx \alpha$ for readily sorbed

TABLE 1. Relative Retention Times in the Analysis of Nonpolar Compounds

Compound	Components in equal concentrations						Trace components					
	Run number					Mean relative t_r	Run number					Mean relative t_r
	1	2	3	4	5		1	2	3	4	5	
n-Pentane	1.00	1.00	1.00	1.00	1.00		1.00	1.00	1.00	1.00	1.00	
n-Hexane	2.00	1.93	2.04	1.98	2.00	1.99	1.98	2.00	2.00	1.99	1.98	1.99
Benzene	2.96	2.96	3.00	2.96	2.98	2.97	2.94	3.02	3.00	3.10	2.92	2.98
n-Octane	10.1	9.9	10.1	9.9	10.1	10.0	10.0	9.9	10.3	9.8	10.1	10.0
Ethylcyclohexane	14.3	14.7	14.9	14.8	15.0	14.7	14.0	14.5	14.6	14.0	14.4	14.3
2,6-Dimethyl-hept-1-ene	16.3	16.4	17.0	16.4	16.6	16.5	15.9	16.5	16.5	16.4	16.2	16.3
n-Nonane	23.8	24.2	25.3	24.0	24.6	24.4	23.2	24.2	23.5	23.5	23.9	23.7

compounds, and so

$$K_c \approx \frac{\alpha - 1}{\alpha + 1}.$$ (6)

An attempt has been made [9] to elucidate how far these quantities are dependent on the concentration of a nonpolar compound in the mixture via a study of the effects of component concentrations on t_r and μ for fast and slow impurities.

Two hydrocarbon mixtures were examined. The first had cyclohexane as base with $10^{-3}\%$ concentrations of impurities, while in the second the hydrocarbons were present in comparable concentrations (about 10%). The trace concentrations were examined with a Panchromatograph system on the 10^{-9} A detector scale, while the ordinary mixture was examined on the 10^{-6} A scale. The samples were always 2 μl used at 75°C with columns 270 × 0.4 cm filled with celite 545 bearing 10% squalane.

First the effects of the nonpolar base zone were examined as regards the relative t_r of the nonpolar impurities (Table 1). It was found that the t_r were almost unaltered (probability of coincidence over 0.8-0.9), so that tables of t_r can be used in identifying nonpolar impurities provided that these were compiled for mixtures with roughly the same component concentrations. This is not unexpected because it has been shown [10] that the retained volume V_N increases linearly with sample size in accordance with

$$V_N = V_R + \frac{q}{2},$$ (7)

where q is the volume of the gaseous sample. Then an appreciable change in V_N (e.g., by 3%) is to be expected for $q = 0.06 V_R$; if $V_R = 200$ cm^3, then q = 12 cm^3 of gas. Then a sample size not exceeding 1-10 μl of liquid or 1-10 ml of gas is hardly likely to affect V_N.

The above estimate is true only if we can neglect adsorption of the trace components at the gas–liquid and (especially) liquid–carrier interfaces, i.e., if the separation occurs under conditions of pure gas–liquid chromatography. If the sorbent is somewhat active in adsorption for the compounds (e.g., in the analysis of nonpolar substances on polar immobile phases or vice versa), the t_r will, in general, be dependent on the concentrations in the sam-

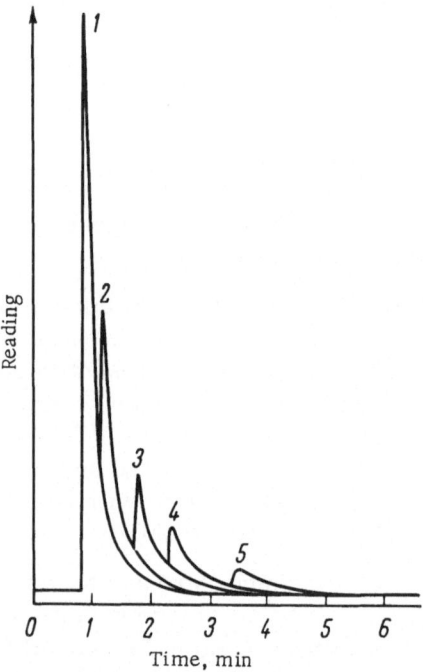

Fig. 2. Reduction in t_r as the ethanol sample size is increased. Sample volumes (μl):
1) 0.70, 2) 0.40, 3) 0.25, 4) 0.15, 5) 0.10.

ple, on the frequency of injection into the column, on the nature of
the other components (especially the base), etc.

If we neglect interaction between components, the retention as
a function of concentration should resemble the relation of V_N to
sample size, which has been examined [11-21] for various carriers, compounds, and liquid phases, and also as affected by sample
size and composition; V_N is usually especially markedly dependent
on q (concentration) for polar compounds separated on nonpolar
phases held on diatomaceous carriers. Figure 2 [22] shows curves
for ethanol on 6% squalane on Spherochrome-1 with various sample sizes; t_r is clearly dependent on sample size and changes by
over 300% as the sample is reduced from 0.7 to 0.1 μl.

A study has been made [23] of the effects of adsorption of
acetone (polar) on V_N for solid carriers impregnated with dinonyl

phthalate with the sample size varying from 1 μg to 1 mg. The carriers were Chromosorb P, celite, teflon 6, and carrier M treated with hexamethyldisilazane; only the last gave a constant specific retained volume as the sample size and content of immobile phase varied.

A relation has been derived [21] between V_N and q; V_N in gas–liquid chromatography is [24] governed by the following equation because the compound dissolves in the liquid and is adsorbed at the two interfaces:

$$V_N(C) - \frac{\partial c_l}{\partial c_g} v_l + \frac{\partial c_{lg}}{\partial c_g} S_l + \frac{\partial c_l}{\partial c_g} \cdot \frac{\partial c_s}{\partial c_l} S_s, \qquad (8)$$

where c_g is the concentration in the gas phase, c_l is the concentration in the liquid phase, c_{lg} is the concentration at the surface of the liquid phase, c_s is the concentration at the surface of the solid, v_l is the volume of liquid in the column, S_l is the surface area of the liquid, and S_s is the surface area of the solid. The compounds usually give isotherms obeying Henry's equation:

$$c_l = K_l c_g. \qquad (9)$$

Then the relative V_N (i.e., V_N relative to the volume of the substance taken as standard) is

$$V_{rel} = \frac{V_N}{V_{N st}} = \frac{K_l}{K_{l\,st}} \left[\frac{1 + \dfrac{AS_l}{K_l v_l} + \dfrac{BS_l}{v_l}}{1 + \dfrac{A_{st}S_l}{K_l v_l} + \dfrac{B_{st}S_s}{v_l}} \right], \qquad (10)$$

where A = $\partial c_{lg}/\partial c_g$ and B = $\partial c_s/\partial c_l$. If the adsorption at the interfaces is small relative to the absorption in the liquid, (10) becomes

$$V_{rel} = K_l/K_{l\,st}. \qquad (11)$$

It follows from (10) that, if adsorption makes a substantial contribution to V_N, one has to standardize the solid and the liquid in order to obtain reproducible V_N; but then gas–liquid chromatography loses one of its main advantages, i.e., simple use of columns with reproducible V_{rel}. This explains the tendency to use separation in the pure gas–liquid region, where V_N is dependent only on the solubility in the liquid.

Inertness in the solid is even more important in trace analysis, since the solid surface may produce reversible or irreversible adsorption as well as chemical (catalytic) reactions. Reversible nonlinear adsorption on the solid can cause t_r or V_N to be dependent on the concentration c.

We consider V_N as a function of c (or of q, which is proportional to this) on the assumption that we can neglect adsorption at the liquid–gas interface, while the absorption in the liquid obeys Henry's law and the adsorption at the liquid–solid interface obeys Freundlich's equation ($\beta < 1$):

$$c_S = \alpha c_l^\beta. \tag{12}$$

Then (10) may be put as

$$V_{\text{rel}} = \frac{V_N}{V_{N\,\text{st}}} = \frac{K_l}{K_{l\,\text{st}}} \left[1 + \frac{\lambda \alpha \beta \kappa_l^{\beta-1} S_S}{v_l} \left(\frac{V_N}{q} \right)^{1-\beta} \right], \tag{13}$$

where $V_{N\,\text{st}} = K_{l\,\text{st}} v_l$ and λ is the coefficient of proportionality in $c_{\max} = \lambda(q/V_N)$ (see [2], for example). This equation implies that V_{rel} increases as q decreases, as S_s increases, and as $(1-\beta)$ increases (greater nonlinearity in adsorption). It is best to put (13) in the following form for experimental test:

$$\log(V_{\text{rel}} - V_{\text{rel il}}) = \log A + (\beta - 1)\log\frac{q}{V_{\text{rel}}}, \tag{14}$$

where $V_{\text{rel il}}$ is the V_{rel} due solely to absorption in the liquid and A is a parameter.

Equation (14) fits closely to the results of [11, 21], and the results for β agree well with generally accepted views on the adsorption of organic compounds by solids.

The actual concentration in trace analysis (sample size for the trace compounds) differs by several orders of magnitude from that in ordinary analysis, so one expects that V_N will sometimes be dependent on the component concentration. In fact, it has been shown [25, 26] that errors can arise if polar compounds are identified from t_r recorded with mixtures at high concentrations. This has to be borne in mind in identifying unknown components of mixtures.

The general interpretation of a chromatogram should be based either on t_r recorded under trace-analysis conditions or on tabulated t_r if it has already been shown that the t_r of the expected (supposed) polar impurities are not concentration-dependent.

Often, the method of identifying trace components can be based on the known adsorption behavior on solids.

In general, the kinetic characteristics of a separation are also concentration-dependent.

If the base has a concentration 10^5-10^7 times those of the impurities, the performance of the system may be affected. Peak widths have been examined as functions of concentration [9], and it has been shown [27] that the width μ has a linear relation to the retention time t:

$$\mu = a + bt, \tag{15}$$

where a and b are constants to be found from experiment. This equation is convenient in the simultaneous examination of performance for fast and slow impurities. Also, an equation such as (15) allows one to calculate μ when it cannot be determined directly (e.g., on account of incomplete separation).

Figure 3 shows μ as a function of t for: I) and II) components in equal concentrations, III) in trace analysis.

The measurements were made with a Panchromatograph (made by Pye, UK), with a flame-ionization detector and a column with 10% squalane on celite 545 with 2 μl samples (curves I and III) and with microsamples (curve II).

Trace components in cyclohexane (curve III) were examined with a detector current of 10^{-9} A, while 10^{-6} A was used for the 2 μl samples of components in equal concentrations, but 10^{-9} A again with microsamples.

Curves I and II (equal concentrations) are almost straight lines, and the microsamples give better separation performance; b in (15) is smaller (curve II). Curve III (trace components) has a

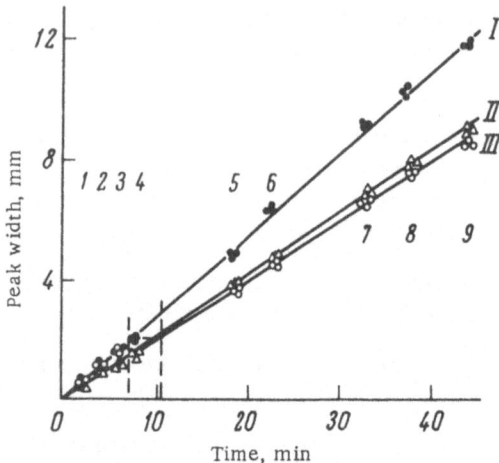

Fig. 3. Relation of peak width to retention time:
I) components present in equal concentrations
2 μl sample; II) the same, microsamples; III)
trace components in a 2 μl sample; 1) pentane;
2) hexane; 3) benzene; 4) cyclohexane; 5) toluene; 6) octane; 7) ethylcyclohexane; 8) 2,6-
dimethylhept-1-ene; 9) nonane. The broken
line shows the width for the base (cyclohexane)
in trace analysis.

pronounced knee near the base zone, and compounds eluted after
the cyclohexane give narrower peaks, so *b* takes one value for the
fast group and another for the slow one. Trace analysis does not
result in additional peak broadening for the heavy components, and
so the base does not affect the performance for any component
whose peak is quite separate from that of the base. This result
does not agree entirely with the common view that, other things
being equal, the separation is better for impurities that run faster
than the base.

A feature of trace analysis is that the base produces a very
broad peak, usually with a very broad tail, which may overlap or
mask the bands for the slow components, or may make these difficult to determine. It is therefore of interest to establish why the
base produces a broad unsymmetrical peak.

A study has been made [28] of the base peak broadening at concentrations of 10^{-3} to 10^{-6} of the maximum concentration. The experiments were done with a Panchromatograph on 270×0.4 cm glass columns with 10% squalane on celite 545 at 75°. Figure 4 shows the peaks given by toluene for fixed 2 μl samples at different detector sensitivities; the sensitivity clearly has marked effects on the width and shape. We might say that we examine the base peak under the microscope when it is recorded at the sensitivity used for the trace components.

An explanation has been offered [28] for the broadening at low concentrations near the baseline in terms of the operative in ordinary chromatographic analysis, though the relation between them is different.

Broadening is usually ascribed [1-3, 50] to diffusion and kinetic factors. Longitudinal diffusion broadening arises from molecular diffusion and eddies; it produces symmetrical broadening of an initially narrow band. Although the initial widths are the same for all components, the maximum concentration in the base band exceeds that for the trace bands by factors of 10^5-10^8, and this means a much larger longitudinal concentration gradient, which from Fick's first law implies that the base band should broaden much more than the others.

Increased longitudinal diffusion can explain the increased width but not the asymmetry (Fig. 4), which is usually ascribed

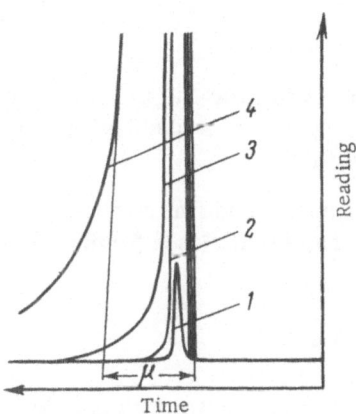

Fig. 4. Effects of detector sensitivity on the peak shape given by 2 μl of toluene; scales (A): 1) 10^{-6}, 2) 10^{-7}, 3) 10^{-8}, 4) 10^{-9}.

to nonlinearity in the partition isotherm, which actually can occur, especially for polar major components on a nonpolar liquid supported on a carrier active in adsorption. If the nonlinear isotherm is convex (e.g., Langmuir isotherm), we can [24] put

$$V_N = K_l v_l + \frac{K_l z p S_S}{(1 + K_l p c_S)^2} ,\tag{16}$$

where $c_S = zpc_l/(1 + pc_l)$ and z and p are the Langmuir parameters. It follows from (16) that V_N is inversely related to the component concentration, i.e., the base peak should have an extended trailing edge.

However, the peak may also be unsymmetrical when the partition isotherm is linear; e.g., the toluene peak in Fig. 4 is clearly unsymmetrical, although the t_r are the same for samples of different sizes with detector currents of 10^{-6} to 10^{-9} A, which indicates that the partition isotherm is linear. A restricted desorption rate is responsible for the skewness in this case.

The carrier gas moves the sample along the column, so the gas concentration is less than the equilibrium one in the tail region [29], which leads to progressive tail broadening. There are thus various reasons for the peak skewness: slow desorption from sorbent grains, slow desorption from stagnant areas, etc.

The base-peak broadening may [28] be evaluated as $\overline{H}$, the height of the equivalent theoretical plate:

$$\overline{H} = \frac{L}{16} \left(\frac{\mu}{t} \right)^2 ,\tag{17}$$

where L is column length, t is retention time for the main component, and μ is the width as defined in Fig. 4. It is usually impossible to assay an impurity if it falls within μ.

Figure 5 shows $\overline{H}$ as a function of gas speed for nitrogen and helium. The $\overline{H} = f(\mu)$ curves have a shape that fits a formula generally used in gas chromatography:

$$\overline{H} = \overline{A} + \overline{B}/u + \overline{C}u ,\tag{18}$$

where $\overline{A}$ represents eddy diffusion, $\overline{B}/u$ represents the contribu-

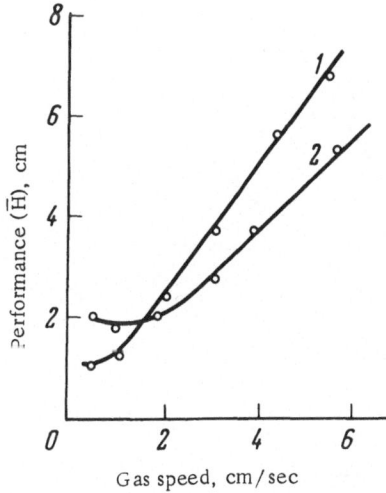

Fig. 5. Effective height H of the equivalent theoretical plate for the main component as a function of gas speed: 1) nitrogen; 2) helium.

tion from longitudinal molecular diffusion, and $\overline{C}u$ represents
mass-transfer resistance. It is no accident that the relationship
is the same in ordinary analysis for the major component as it is
in trace analysis, for the same basic causes of broadening are
operative in both instances. From (18) we get the following re-
sults in trace analysis for nitrogen: $\overline{A}$ = 0.05 cm, $\overline{B}$ = 0.19 cm^2/sec,
and $\overline{C}$ = 1.15 sec, while for helium $\overline{A}$ = 0.05 cm, $\overline{B}$ = 0.69 cm^2/sec,
and $\overline{C}$ = 1.00 sec. The corresponding results for ordinary analysis
are as follows: nitrogen A = 0.048 cm, B = 0.014 cm^2/sec, and C =
0.009, helium A = 0.057 cm, B = 0.02 cm^2/sec, C = 0.005 sec.

The main parameters governing the broadening are thus C and
A at low concentrations. Also, the width is linearly related to
sorbent particle diameter, while $\overline{H}$ for ethanol is reduced (Fig. 6)
when the carrier is modified by 1% of polar triethanolamine. This
shows that kinetic factors play a major part in producing the width
of the band for the main component. Further theoretical and experimental research is needed on base-peak broadening at low
concentrations, as this is important not only to trace analysis but
also to preparative gas chromatography.

It is important to estimate how the separation is affected by
the working parameters in order to choose the best conditions for

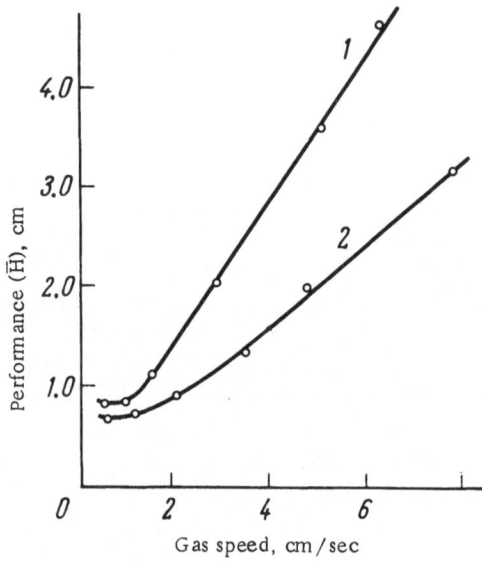

Fig. 6. Effective height $\bar{H}$ before (1) and after (2)
carrier modification by triethanolamine. 270×0.4
cm column, 12% squalane on Spherochrome 1,
75°C, 2 μl sample.

separating trace components from major ones; it has been shown
[4, 30, 31] that the number of theoretical plates needed for a given
separation increases considerably with m = $\bar{c}_{max}/c_{max}$ ($\bar{c}_{max}$ is
the maximum concentration for the base and c_{max} is the same for
an impurity). It is found [31] that 0.6% of impurity can be observed
if the two components differ in retention time by three standard deviations, while 10^{-3}% can be detected if the difference is four standard deviations. These studies involved the assumption that the
main peak had a gaussian form; but that peak is usually unsymmetrical for m large, and it does not have that shape.

Often, an impurity peak appears on the elongated tail of the
main peak (with a shifting baseline), which is due to a monotonic-
ally changing concentration of the main component. The observed
curve then has the impurity peak (gaussian curve) superimposed
on the falling tail. A chromatogram from an ordinary mixture differs from that for a trace analysis, so special semiempirical cri-
teria have been suggested [32–34] for peak separation in the trace

case, of which the ψ proposed by Vigdergauz et al. [35] appears
the best:

$$\psi = \frac{h_i - h_{min}}{h_i}, \tag{19}$$

where h_i is impurity peak height and h_{min} is the height of the minimum. Here the heights are reckoned from the baseline, which is
a disadvantage, because the value can vary greatly in trace analyses that (from the viewpoint of the analyst) involve identical separations. It would thus appear better to use a quantity $\bar{\psi}$ (Fig. 7):

$$\bar{\psi} = \frac{f}{g}. \tag{20}$$

Here f is the ordinate of the maximum for the impurity, as reckoned from the level of the minimum, while h_i is replaced by the
true peak height g, which is reckoned from the background (baseline) level at the impurity maximum. If there is no minimum
(merely an inflection), $\bar{\psi}$ becomes negative. This $\bar{\psi}$ characterizes
not so much the separation as the visibility of the impurity peak
on the tail from the major component.

   If the impurity peak is considered approximately as an isos-
celes triangle (the basis of one method of calculating peak area),
then h_i and the width $\mu_{1/2}$ at half height are related [36] by theo-
retical-plate theory as follows:

$$h_i = P_c \frac{\sqrt{N}}{V_R} \cdot \frac{q}{\sqrt{2\pi}} = P_c c_{max}, \tag{21}$$

$$\mu_{1/2} = 4P_t \frac{V_R}{\sqrt{N}}, \tag{22}$$

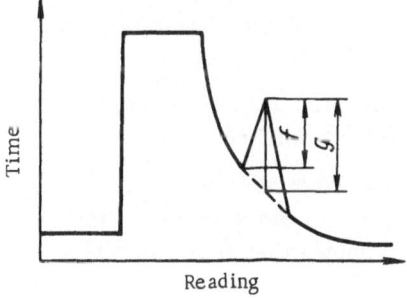

Fig. 7. Definition of the separation criterion $\bar{\psi}$ for an impurity and the major
component.

where q is the total amount of the impurity in the sample and P_c and P_t are coefficients of proportionality. The concentration curve for the base component can be considered roughly as a straight line with the following slope near the impurity peak:

$$K_0 = \frac{P_c}{P_t} \left(\frac{\partial c_b}{\partial V} \right) V_{R_i}, \tag{23}$$

where c_b is the base concentration and V_{R_i} is the retention volume for the impurity. From (21)-(23) we can express $\bar{\psi}$ in terms of the parameters of the chromatographic run:

$$\bar{\psi} = \frac{f}{g} = \frac{\mu_{1/2}(K_i + K_0)}{\mu_{1/2} K_i} = 1 + \frac{K_0}{K_i}, \tag{24}$$

where K_i is the slope of the impurity peak in the absence of the base. As

$$K_i = \frac{h_i}{\mu_{1/2}} = \frac{P_c}{P_t \sqrt{8\pi}} \cdot \frac{qN}{V_R^2}, \tag{25}$$

the separation criterion becomes

$$\bar{\psi} = 1 + P_i \frac{K_0 V_R^2}{Nq}, \tag{26}$$

where P_i is a coefficient of proportionality.

We see from (26) that the separation is complete if $\bar{\psi} = 1$ (and $K_0 = 0$); $\bar{\psi} < 1$ for incomplete separation because K_0 is negative.

The separation improves as the absolute magnitude of the second term in (26) decreases, i.e., with the column performance, with q, and as K_0 and V_R decrease. The best conditions are those that make $\bar{\psi}$ as large as possible.

It is of interest to examine how $\bar{\psi}$ varies with the parameters, since this influences the choice of the best conditions.

It has been found that $\bar{\psi}$ is almost independent of q in the range 0.5-7 μl, but the dependence on gas speed is more complicated, as this speed affects K_0 and N. Figure 8 shows that there is a best speed, which lies in the region optimal for the column under ordi-

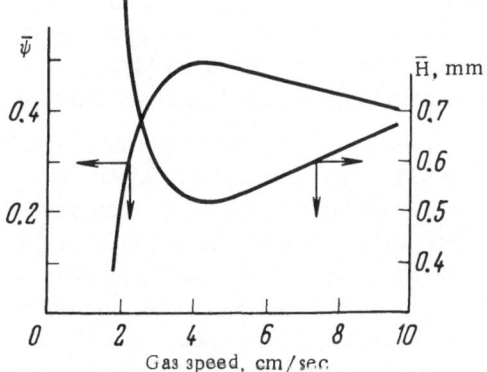

Fig. 8. Dependence of $\bar{\psi}$ and $\bar{H}$ for octane (impurity) in benzene on gas speed. 270×0.4 cm column, 10% squalane on Spherochrome 1, 75°C, 2 µl sample.

nary analysis conditions (compare the curve for H as a function of the speed).

This elucidation of separation in trace analysis provides a better basis for the choice of the best conditions and for the development of methods.

One can get irreversible adsorption or catalytic reactions in separations in gas-adsorption chromatography and gas—liquid chromatography, and these can lead to serious errors in interpreting analyses, especially trace ones.

For instance, only 10^{-4} to 10^{-6} of the base may react during a separation, which will not affect the results if the components are present in comparable concentrations; but even a small degree of reaction can produce false peaks in a trace analysis, so especial attention has to be given to the catalytic activity of the sorbent in relation to the major component. It is possible to distinguish these false peaks because their heights increase with the duration of the run.

The nature of the solid is usually decisive for the catalytic activity. For instance, decomposition of methyl cis-9,10-epoxyoctadecanoate at 235°C on a column bearing silicone oil can be avoided if one uses the more inert celite instead of C-22 refractory brick

[37]. Spencer [38] notes that alcohols, amines, and organic acids can be adsorbed on solid carriers, while pesticides and steroids decompose on metal surfaces. See also [3, 39, 40] on the catalytic activities of solid carriers in gas–liquid chromatography.

Irreversible adsorption can be the source of major errors in separations; Kusy [41] made the first detailed study of this, although adsorption by the carrier had been reported before [42, 43]. Direct evidence for adsorption has been given [44] from the use of tagged compounds (3% benzene and acetone containing ^{14}C in cyclohexane). The tests were done at 85°C with squalane on celite 545, INZ-600, Spherochrome 1, and the latter modified with 1% triethylene glycol. The recordings were made with a katharometer and with a radiation detector.

The adsorption was examined in two ways. In the first, a relatively large sample of the same compound (inactive form) was injected as soon as the peak from a tagged component had been recorded. If the adsorption energy was not too large, one would expect that the inactive form would exchange with or displace the adsorbed tagged form, and so a repeat analysis for the inactive compound should yield a radioactive peak. This was so for the polar acetone.

In the second method, the gas flow was stopped as soon as the peak for the tagged compound had been recorded; the column was taken out of the instrument, the sorbent was removed, the liquid phase was dissolved in acetone, and part of the solution was used in a radiation measurement, which gave an estimate of the adsorption.

It was found that none of the carriers gave appreciable adsorption of the nonpolar benzene, but there was marked adsorption of the polar acetone. The various methods gave from 2.2 to 5.6% less of the sample to a column of Spherochrome 1. Modification with triethylene glycol reduced this to 0.6%.

This is direct evidence for adsorption of polar compounds by the carrier in gas–liquid chromatography; the nature of the sorbent and of any modifier influence the uptake considerably.

Kusy's study [41] was restricted to ordinary mixtures; the trace region was not considered, although irreversible adsorption should affect quantitative results particularly markedly in that case.

Berezkin et al. [45] examined the effects of adsorption on quantitative results in trace analysis by the use of carriers treated with 10% of Apiezon N. The initial carrier was Spherochrome 1 (0.16-0.25 mm fraction); some runs used carriers treated with dimethyldichlorosilane or 1% triethylene glycol. A KhV-2 chromatograph was used with 1% component concentrations, while a Tsvet-1 was used for $5 \cdot 10^{-3}\%$. The 200×0.4 cm columns were operated at 90°C with helium flowing at 50 ml/min and samples of 2-5 μl.

Table 2 gives the results in terms of $Z_i = S_i/S_{st}$, where S_i is the peak area for component i and S_{st} is the area for the standard (n-nonane), the latter being chosen because it should give slight and unspecific adsorption on diatomite [46].The table shows that toluene (low polarity) gave practically identical results in all mixtures, while the more polar methyl ethyl ketone gave Z_i reduced by adsorption when unmodified carriers were used instead of modified ones. For instance, Z_i fell by 18% on going from carrier III to carrier II with mixture 1, and by 28% on going to carrier I, which was due to virtually irreversible adsorption of the methyl ethyl

TABLE 2. Quantitative Analyses for Polar and Nonpolar Impurities with Various Carriers

Mixture	Main component	Concentration, %	Impurity	Z_i		
				I	II	III
				Unmodified carrier	Treated with methyl-dichloro-silane	Treated with 1% tri-ethylene glycol
1	Heptane	1.0	Toluene	1.08	1.16	1.12
			Methyl ethyl ketone	0.78	0.89	1.09
			n-Amyl alcohol	—	—	0.76
			n-Nonane	1.0	1.0	1.0
2	Methanol	1.0	Toluene	1.13	1.12	1.13
			Methyl ethyl ketone	0.93	0.93	1.12
			n-Amyl alcohol	—	—	0.78
			n-Nonane	1.0	1.0	1.0
3	Cyclo-hexane	$5 \cdot 10^{-3}$	Toluene	0.72	0.74	0.75
			Methyl ethyl ketone	—	0.91	1.15
			n-Amyl alcohol	—	—	0.65
			n-Nonane	1.0	1.0	1.0

ketone. Mixture 2 had the polar base (methanol) running ahead of the ketone, and here the results for carriers I and II were higher than those for mixture 1, where the base was nonpolar heptane. This was due to modification and displacement by the polar methanol. A polar base can sometimes produce spurious peaks by displacing components adsorbed in previous analyses (column memory). The most polar impurity used (amyl alcohol) was eluted quantitatively only from the modified carrier III.

Mixture 3 ($5 \cdot 10^{-3}\%$ impurities) gave the same result for toluene (nonpolar) on all three carriers; but Z_i for the ketone fell by 21% between III and II, and it could not be assayed with the unmodified carrier. The more polar n-hexanol was eluted only from carrier III. Adsorption thus has more effect as the impurity concentration decreases, as one would expect. The μ for methanol (as base) increased by a factor 1.6 between carriers III and II, and by 2.4 for carrier I. Good results can be obtained in impurity analysis if the carrier is treated with small amounts of polar compounds [47]. Methods devised for impurity analysis should also be checked against standard mixtures whose compositions are close to those of the samples, since separation of all components is a necessary but not sufficient condition for correct trace analysis.

If corrosive polar compounds are to be analyzed and these are adsorbed by the carrier or apparatus, or perhaps react with the stationary liquid, it is useful to condition the column by periodically injecting the mixture over a prolonged period [48].

The above features must be borne in mind in developing particular methods of analysis.

LITERATURE CITED

1. A. J. M. Keulemans, Gas Chromatography, 2nd edition, edited by C. G. Verver, New York, Reinhold (1959).
2. A. A. Zhukhovitskii and N. M. Turkel'taub, Gas Chromatography [in Russian], Moscow, Gostoptekhizdat (1962).
3. S. D. Nogare and R. S. Juvet, Gas—Liquid Chromatography, Theory and Practice (1962).
4. E. Glückauf, Trans. Faraday Soc., 51:34 (1955).
5. W. L. Jones and K. Kiselbach, Anal. Chem., 30:1590 (1958).
6. Pure and Appl. Chem., 1:177 (1960).

7. N. M. Turkel'taub and A. A. Zhukhovitskii, Gas Chromatography, Proceedings of the First All-Union Conference [in Russian], Moscow, Izd. AN SSSR (1960), p. 144.

8. A. A. Zhukhovitskii and N. M. Turkel'taub, Usp. Khim., 30:877 (1961).

9. V. G. Berezkin, V. S. Tatarinskii, and L. L. Starobinets, Zh. Anal. Khim., 24:600 (1969).

10. P. E. Porter, C. H. Deal, and F. H. Stross, J. Amer. Chem. Soc., 78:2999 (1956).

11. P. G. Scholz and W. W. Brandt, Gas Chromatography, edited by N. Brenner, J. E. Callin, and M. D. Weiss, New York, Academic Press (1962), p. 7.

12. J. J. Kirkland, Anal. Chem., 35:2003 (1963).

13. P. Urone and J. F. Parcher, Anal. Chem., 38:270 (1966).

14. T. Fukude, Japan Analyst, 8:627 (1959).

15. E. B. Bens, Anal. Chem., 33:178 (1961).

16. R. L. Pecsok, A. Yllana, and A. Abdul-Karim, Anal. Chem., 36:452 (1964).

17. I. A. Musaev, P. I. Sanin, V. P. Pakhomov, V. G. Berezkin, N. N. Barinova, and D. K. Zhestkov, Neftekhimiya, 6:131 (1966).

18. T. B. Gavrilova and A. V. Kiselev, Gas Chromatography, Proceedings of the Third All-Union Conference [in Russian], Izd. Dzerzh. Fil. OKBA (1966), p. 204.

19. V. G. Berezkin, Yu. A. Kolbanovskii, and É. A. Kyazimov, Zh. Fiz. Khim., 40:1921 (1966).

20. E. Cremer and F. Prior, Z. Electrochem., 55:217 (1951).

21. V. G. Berezkin and V. P. Pakhomov, Zh. Fiz. Khim., 42:1849 (1968).

22. V. G. Berezkin, V. P. Pakhomov, and V. R. Alishoev, Gas Chromatography [in Russian], No. 11, Moscow, NIITÉKhim (1969).

23. N. A. Cockle and P. F. Tetley, Chem. Ind., 1118 (1968).

24. V. G. Berezkin, V. P. Pakhomov, V. M. Fateeva, and V. S. Tatarinskii, Dokl. AN SSSR, 180:119 (1968).

25. M. J. E. Golay, Nature, 202:489 (1964).

26. R. G. Ackman, J. Gas Chromat., 3:15 (1965).

27. B. D. Blanstein and J. M. Heldman, Anal. Chem., 36:65 (1964).

28. V. G. Berezkin and V. S. Tatarinskii, Izv. AN SSSR, Ser. Khim., No. 7 (1970).

29. G. Schay, Theoretical Principles of Gas Chromatography [Russian translation], Moscow, IL (1963).

30. A. A. Zhukhovitskii and N. M. Turkel'taub, Neftekhimiya, 3:135 (1963).

31. A. N. Genkin, Physicochemical Methods of Analysis and Examination of Synthetic-Rubber Products [in Russian], Leningrad, Goskhizdat (1961), p. 55.

32. H. G. Struppe Gas-Chromatographie 1958, Berlin, Akademie-Verlag (1959), p. 528.

33. K. Kauser, Gas-Chromatographie, Leipzig (1960).

34. H. Röck, Chem. Ing. Techn., 28:489 (1956).

35. M. S. Vigderganz, M. I. Afanas'ev, and K. A. Gol'bert, Usp. Khim., 32:754 (1963).

36. A. J. P. Martin and R. L. Synge, Biochem. J., 35:1358 (1941).

37. A. P. Tulloch, B. M. Craig, and G. A. Ledingham, Canad. J. Microbiol., 5:485 (1959).

38. S. F. Spencer, Instrumentation in Gas Chromatography, Eindhoven, Centrex Publishing Company (1967), p. 131.

39. H. P. Burchfield and Eleanor E. Storrs, Biochemical Applications of Gas Chromatography, New York, Academic Press (1962).

40. V. G. Berezkin and V. P. Pakhomov, Izv. AN SSSR, Ser. Khim., 791 (1968).

41. V. Kusy, Anal. Chem., 37:1748 (1965).

42. A. Karmen, L. Ginffrida, and R. L. Bowman, J. Lipid Res., 3:44 (1962).

43. H. Bührins, J. Chromat., 11:452 (1963).

44. V. S. Tatarinskii, V. G. Berezkin, A. A. Efremov, Ya. D. Zel'venskii, L. N. Kolomiets, and V. I. Morozov, Izv. AN SSSR, Ser. Khim., 2634 (1968).

45. V. G. Berezkin, V. P. Pakhomov, and V. R. Alishoev, Khim. i Tekh. Topl. i Masel, 56 (1967).

46. A. V. Kiselev and Ya. I. Yashin, Gas-Adsorption Chromatography [in Russian], Moscow, Nauka (1967).

47. H. A. Saraff, A. Karmen, and J. W. Heavy, J. Chromat., 9:122 (1962).

48. Yu. S. Drugov, Ph.D. Thesis, Institute of Industrial Hygiene and Occupational Diseases, Academy of Medical Sciences of the USSR, Moscow (1967).

49. L. S. Ettre and A. Zlatkis, The Practice of Gas Chromatography, New York, Interscience (1967).

50. J. C. Giddings, Dynamics of Chromatography, part I, London, Edward Arnold (Publ.) Ltd., New York, Marcel Dekker Inc. (1965).

Chapter II

Use of Large Samples in Impurity Analysis

The bands broaden continuously in a column during development, and the peak concentration falls, which restricts the limit of detection, since an impurity can be indicated by a concentration-sensitive detector only provided that the peak concentration is greater than the minimum detectable concentration c_d:

$$c_{max} \geqslant c_d. \tag{27}$$

A flux detector responds to flows greater than j_d, and the recording condition may accordingly be put as

$$c_{max} F \geqslant j_d, \tag{28}$$

where F is the flow speed. If (27) or (28) is not met, the detector does not record the impurity.

The conditions must be appropriately altered in order to increase the concentration or flux. Theory [1, 2] indicates that c_{max} increases with N and q, and also as V_R decreases:

$$c_{max} = \frac{N^{1/2}}{V_R} \cdot \frac{q}{(2\pi)^{1/2}}. \tag{29}$$

The simplest method is therefore to increase q; but this increases the width and reduces N, which results in poorer separation, and so this restricts the use of large samples in impurity analysis.

The problem in any particular case is to find the best com-
promise between sensitivity and separation in adjusting q. The
best q is determined by the detailed conditions, but in each particular case one has to take account of the general effects of q
(reduced column performance but increased c_{max} as q is increased).

Here we consider performance as a function of q on the assumption that we can neglect the effects of the base (see Chapter I
on separation of impurities from the base).

There are [3-6] theoretical studies of peak shape as a function
of q. Kalmanovskii and Zhukovitskii [7, 8] devised novel calculation methods of considerable interest to the choice of best conditions. There are also other studies [9-12] of column performance
in relation to q.

If the sample takes the form of a concentration rectangle c_0
(piston method), increase in sample size (piston width) causes increases in peak width and in c_{max}. Figure 9 [8] shows the general
form of the output curves in relation to sample size in this case.
As characteristics we use B (the time width of the initial peak),
c_0, c_{max}, τ_0 (peak width at half height for a vanishingly small sample), and τ (width corresponding to the actual sample $q = uBc_0$,

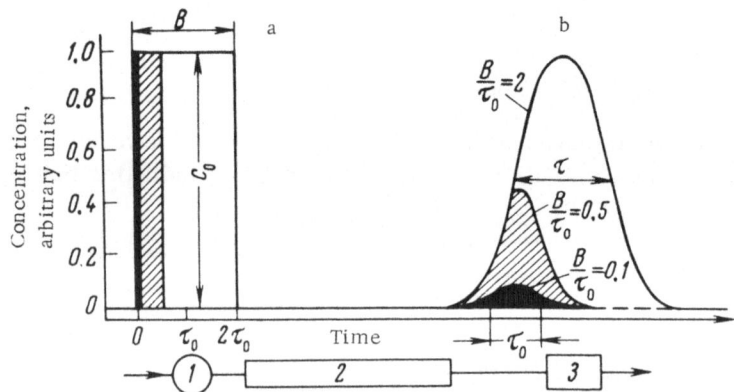

Fig. 9. Effects of sample size (a) on peak shape (b) in the piston
method [8]: 1) sample injection, 2) chromatographic column,
3) detector.

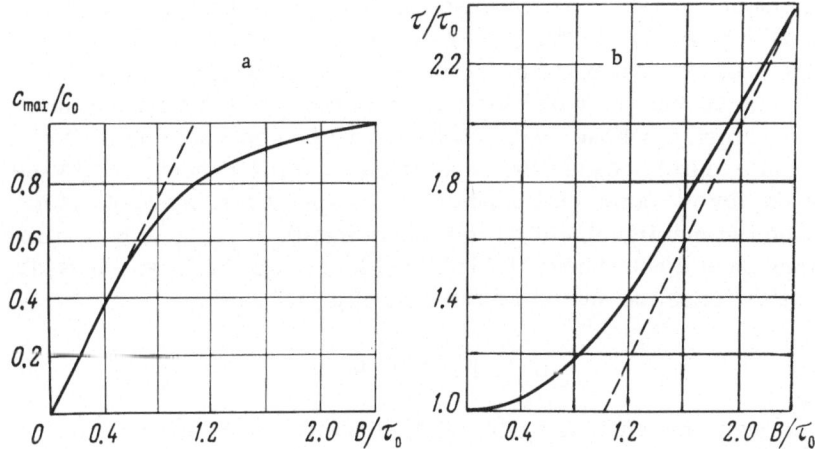

Fig. 10. a) Relative peak concentration and b) relative peak width as functions of relative sample size B/τ_0 [8].

where u is gas speed). Figure 10 [8] shows in dimensionless terms the results for c_{max}/c_0 and τ/τ_0 as functions of B/τ_0.

We see that c_{max}/c_0 increases with B/τ_0 nearly linearly up to 0.8, at which point the relative width at half height exceeds τ_0 by only 15%, which often represents a negligible deterioration in separation. Here $c_{max} = 0.65c_0$, so one is not justified in adopting the conventional view that the largest permissible sample size q_{max} should be so small that τ is governed solely by column broadening and is independent of initial sample size [8]. That approach leads to an unjustified loss in sensitivity.

The results of Fig. 10 allow one to estimate q_{max} for particular working conditions.

It is best for an overloaded column to represent V_a (initial volume of a rectangular sample, sample width μ_s) as the sum of individual narrow rectangular (unit) bands, each having the volume V_s (the limitingly small volume of a narrow sample, which produces an output width independent of sample size). Then there are $n = V_a/V_s$ such unit zones. V_s produces a bell-shaped unit output peak, whose base width is μ_b (in volume units). If one compound does not interfere with another, and if there is no effect from the

injection sequence, we can use the superposition principle to dis-
cuss approximately the band given by an overloaded column as the
result of summation of n unit bell-shaped bands representing V_a.
We can deduce the width via the processes for the first and last
unit samples, whose peak positions will be separated by μ_s (in
volume units) because the separations are independent, while the
peaks for all other unit bands from 2 to $n-1$ fall between them.
Then the maximum width from the points defining μ_s in both direc-
tions should not exceed the half-width of a unit band, so the width
μ at the baseline should be the sum of μ_s and μ_b:

$$\mu = \mu_s + \mu_b. \tag{30}$$

This equation enables us to relate directly the relative loss of column performance to sample size (or initial width):

$$\frac{H}{H_b} = \frac{N_b}{N} = \left(1 + \frac{\mu_s}{\mu_b}\right)^2, \tag{31}$$

where H and H_b are heights equivalent to a theoretical plate in
separating samples having volumes V_a and V_s, respectively, while
N and N_b are the numbers of theoretical plates applicable to those
volumes.

Equation (31) is a general relation of column performance to
sample size and applies over a wide range in size [13]. Figure 11

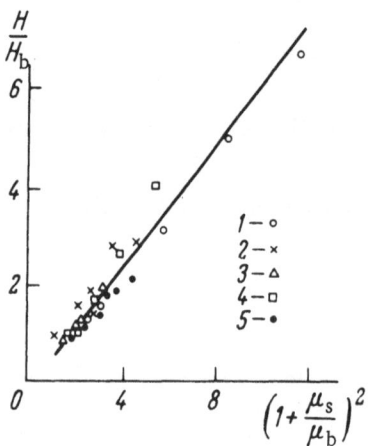

Fig. 11. Dependence of H/H_b on
$(1 + \mu_s/\mu_b)^2$. Carrier gas 40 ml/min,
silica gel treated with 1.5% vaseline
oil: 1) ethane 470 × 0.4 cm column;
2) propane, 470 × 0.4 cm; 3) propane,
320 × 0.4 cm; 4) propane, 320 × 0.4
cm; 5) from [9].

gives the relative changes in performance for various speeds, sample sizes, and compounds; it includes results from other sources [9]. All the experimental results agree satisfactorily with (31).

Equation (31) also agrees with the results of Hollingshead et al. [14]; they showed that pulse injection gave the same relation of H/H_b to sample volume for a wide range of compounds and working conditions. The above equation also yields a useful relation for graphical derivation of the equivalent height of a theoretical plate with a very small sample:

$$H'^{1/2} = H_b^{1/2} + \lambda B. \tag{32}$$

The relation of H/H_b to run parameters for a constant sample size has also been discussed.

The relative deterioration in component separation can be deduced from

$$\frac{R}{R_b} = \left(1 + \frac{\mu_s}{\mu_b} \right)^{-1}. \tag{33}$$

It follows from (32) and (33) that the performance and separation should deteriorate as the sample size increases, as is found; the increase in H (reduction in performance) is more pronounced for compounds having small V_R [14]. We get $H(V_H)$ directly from (34), which is obtained by transforming (32):

$$H = \left(\sqrt{H_b} + \frac{4\mu_s}{V_N} \right)^2. \tag{34}$$

The best sample size will thus differ as between fast and slow impurities.

Table 3 gives examples of the basic parameters of some methods employing large samples.

If the sample is large enough, c_{max} approaches c_0; for instance, $c_{max}/c_0 = 0.995$ [9] if $V/\mu_b = 2.4$. If the sample is large, there is not enough time for the initial band to spread out completely, and only the edges spread, so the output curves have a stepped form. This approach was suggested independently by

TABLE 3. Basic Parameters of Some Chromatographic Methods of Impurity Analysis via Large Sample with Katharmometer Detection

Major component	Gas (ml) or liquid (μl)	Trace components	Limit, wt.% *	Separation conditions			Reference
				T,°C	Column length, m	Sorbent	
Ethylene	25 (gas)	Hydrogen Oxygen Methane Carbon monoxide Nitrogen	$5 \cdot 10^{-4}$ $4 \cdot 10^{-4}$ $7 \cdot 10^{-4}$ 10^{-3} $7 \cdot 10^{-4}$	25	2	13X molecular sieve	[15]
Ethylene	5 (gas)	Propane Propylene C_4 hydrocarbons	$5 \cdot 10^{-3}$ $5 \cdot 10^{-3}$ 10^{-2}	25	8	Triisobutylene on celite	[15]
Acetylene	50-100 (gas)	Methylacetylene, vinylacetylene	$< 10^{-2}$	25-50	1-3.5	Dibutyl phthalate on Inza brick	[16]

Major component	Gas (ml) or liquid (μl)	Trace components	Limit, wt.% *	Separation conditions			Reference
				T,°C	Column length, m	Sorbent	
Propylene	25 (gas)	Propadiene	$5 \cdot 10^{-4}$	25	2	Silver nitrate dissolved in glycol	[17]
Acetylene	30-100 (gas)	Carbon monoxide Methane Carbon dioxide	$1 \cdot 10^{-2}$	80	2.5-3.0	ACM silica gel	[18]
Butadiene	10-50 (gas)	Hydrogen Nitrogen Oxygen Methane Carbon monoxide	$10^{-2}-10^{-4}$	—	—	—	[19]
Benzene	10 (liquid)	Isopropanol	$< 10^{-4}$	64	1.2-2.0	Nujol on celite	[20]
Toluene	10 (liquid)	Cyclohexanol	$4 \cdot 10^{-4}$	153	1.2-2.0	Carbowax and sorbitol on celite	[20]

*The limit of detection is expressed in volume %.

Zhukhovitskii and Turkel'taub [21-23] and by Reilly et al. [24]; it has been called step chromatography [21-23]. If the separation is performed under isothermal conditions, c_0 is the concentration in the flat top, so increase in sample size is desirable only if the initial impurity concentration in the sample exceeds the concentration limit for detection. If this is not so, impurity concentration or chromatothermographic methods must be used.

There are the following advantages in this method:

1. The signal is more stable;
2. The peak (step) height is directly proportional to the concentration, which simplifies calculation and quantitative interpretation;
3. There are no errors due to partially irreversible adsorption;
4. The limit of detection is improved somewhat.

A flat-topped peak is formed only above a certain q defined [25] by

$$q > 3.2K \sqrt{N},$$

where K is the coefficient for partition of the substance between the gas and the sorbent.

The step method has the disadvantage that the flat-topped band is broad, so it is best used in impurity analysis for systems where the relevant components are very well separated. The following is [25] the condition on q for separation of two flat-topped bands:

$$q < L\Delta K - 0.7K \sqrt{N}, \tag{35}$$

where L is column length.

Figure 12 [26] is a step chromatogram illustrating the determination of ethylene and propane as impurities in town gas.

This method has been used by Palamarchuk [27] for assaying dimethyldichlorosilane for traces of trimethyltrichlorosilane and methyltrichlorosilane with a limit of detection of about $5 \cdot 10^{-3}\%$.

Pulsed frontal gas chromatography [28, 29] can be useful for impurity assay in certain cases. If there is no restriction on q, it may be best to use vacant chromatography [30, 31], which was considered independently for certain particular cases by Reilly et al. [24]. In that case the column is supplied with a flow of the analysis mixture, not of carrier gas. If a known volume of a pure component is introduced into this flow ahead of the column, we get vacant regions (regions of reduced concentration), which move along the column with various speeds. The output curve then contains as many peaks as there are components in the sample.

The differential equations for this method show that the dips move and broaden in the same way as bands in ordinary chromatography, so the usual methods of calibration and interpretation still apply. The method has [23] the following advantages:

1. The mixture is passed continuously, and no carrier gas is needed;
2. Sample dispensing is simplified;
3. The sum of the component concentrations is measured continuously;
4. The result for the concentration is not an instantaneous one but an average over a certain period.

The method provides a new means of adjusting the analysis conditions. If the mixture passes first through a short column of adsorbent, we get dips for all components, whose sizes are dependent on the adsorption behavior of the substances.

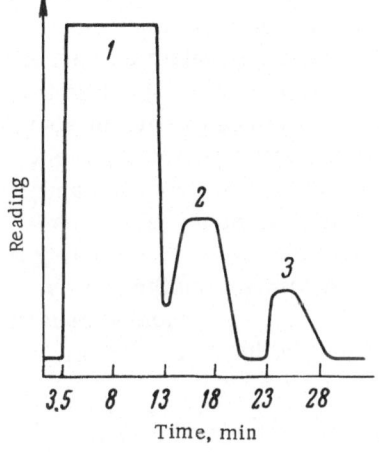

Fig. 12. Step chromatography in impurity determination in town gas: 1) hydrogen, methane, and ethane; 2) ethylene; 3) propane. Separation on a 3-m alumina column.

A selective sorbent produces dips for the sorbed components. For instance, a hydrocarbon mixture may be passed through a sorbent containing bromine to produce dips only for the unsaturated hydrocarbons. An interesting form of the method is of value in impurity analysis and is called differential chromatography [32], in which the column receives a continuous flow of the sample, to which is periodically added a mixture of known composition. The peaks or dips on the curve represent the deviations of the current concentrations from the known ones; no peak or dip is produced if the two mixtures have the same concentration for a particular compound.

Vacancy chromatography is used in impurity analysis on high-purity gases, e.g., by passing the gas through a needle valve and a stopcock with a 5-ml dispensing volume to the 100×0.6 cm column [33], which contains 13X molecular sieve. The dispensing space is evacuated and then is connected into the flow line; impurity fractionation occurs, and dips are produced at the output. The signal amplitude increases as the hole size in the needle valve decreases and as the pressure in the dispensing space is reduced. Dips are also produced when the gas is passed through a capillary 70 cm long and the pressure is suddenly reduced at the inlet to the column, e.g., by piercing a rubber diaphragm with a needle. Analogous methods of producing dips have been described previously [23, 25]. The limit of detection in [33] was about $10^{-4}\%$.

A similar method has been used to detect permanent gases by means of high-sensitivity ionization detectors. A small amount of ethylene or acetylene is passed through the ionization detector after the column, which gives rise to a standing ionization current, which is markedly reduced by the entry of permanent gas into the detector. Willis [34] found that an ethylene concentration of about $10^{-4}\%$ could be used with an argon radioactive detector to give a minimum detectable concentration of about $10^{-4}\%$ for argon and about $0.5 \cdot 10^{-4}\%$ for hydrogen, oxygen, and methane. The background can be produced by various substances, e.g., 1,2,4,5-tetrachlorobenzene [35] or p-toluidine [36]. A similar method can be used with a flame-ionization detector [37, 38]. The method usually provides a sensitivity higher than that available from a katharometer.

Klesment [39] has used hydrogenation and dehydrogenation in the analysis of organic compounds. The compounds are separated chromatographically in a flow of inert gas containing a small amount of hydrogen, and then the flow enters a platinum catalyst, which hydrogenates or dehydrogenates the compounds, which are then completely taken up in a trap containing activated charcoal. The katharometer then measures only changes in the hydrogen concentration.

Recently Klesment [40] has described an analogous method based on catalytic oxidation of organic compounds differing in reactivity; helium containing 4% oxygen was used as carrier. The method can be used in impurity analysis. Here the oxygen acts as an indicator for the organic compound. In principle, any compounds can be used as an indicator in this way, and this provides scope for selective determination of impurities.

LITERATURE CITED

1. A. J. P. Martin and R. L. M. Synge, Biochem. J., 35:1358 (1941).
2. A. J. M. Keulemans, Gas Chromatography, 2nd edition, edited by C. G. Verver, New York, Reinhold (1959).
3. P. E. Porter, C. H. Deal, and F. H. Stross, J. Amer. Chem. Soc., 78:2999 (1956).
4. E. Glückauf, Trans. Faraday Soc., 51:34 (1955).
5. E. Glückauf, Trans. Faraday Soc., 60:729 (1964).
6. J. J. van Deemter, F. J. Zuidenweg, and A. Klinkenberg, Chem. Eng. Sci., 5:271 (1956).
7. V. I. Kalamanovskii and A. A. Zhukhovitskii, J. Chromat., 18:243 (1965).
8. T. B. Gavrilova and A. V. Kiselev, Gas Chromatography, Proceedings of the Third All-Union Conference [in Russian], Izd. Dzerzh. Fil. OKBA (1966), p. 204.
9. H. Purnell and D. T. Sawyer, Anal. Chem., 36:668 (1964).
10. M. S. Vigdergauz and M. I. Afanas'ev, Neftekhimiya, 3:911 (1964).
11. A. A. Zhukhovitskii and N. M. Turkel'taub, Neftekhimiya, 3:135 (1963).
12. A. N. Genkin, Physicochemical Methods of Analysis and Examination of Synthetic-Rubber Products [in Russian], Leningrad, Goskhizdat (1961), p. 55.
13. V. G. Berezkin and O. L. Gorshunov, Zh. Fiz. Khim., 42:2587 (1968).
14. I. W. Hollingshead, H. W. Habgood, and W. E. Harris, Canad. J. Chem., 43:1560 (1965).
15. G. Nodop, Z. Anal. Chem., 164:120 (1958).
16. L. M. Kontorovich, A. V. Iogansen, T. T. Levchenko, G. N. Semina, V. P. Bobrova, and V. A. Stepanova, Zav. Lab., 28:146 (1962).
17. E. Bla, P. Manaresi, and L. Motta, Anal. Chem., 31:1910 (1959).
18. L. M. Kontorovich and V. P. Bobrova, Gas Chromatography, Proceedings of the Second All-Union Conference [in Russian], Moscow, Nauka (1964), p. 150.
19. Lu P'ei-chang, Wang Ts'ai-chu, Wang Jui-hua, and Wang Hsiao-yii, Hua Hsueh Hsuehbao, 32:26 (1966).
20. C. E. Behhet, S. D. Nogare, L. W. Safranski, and C. D. Lewis, Anal. Chem., 30:898 (1958).

21. A. A. Zhukhovitskii and N. M. Turkel'taub, Dokl. AN SSSR, 144:829 (1962).
22. A. A. Zhukhovitskii and N. M. Turkel'taub, Gas Chromatography, Proceedings of the Second All-Union Conference [in Russian], Moscow, Nauka (1964), p. 12.
23. A. A. Zhukhovitskii and N. M. Turkel'taub, Neftekhimiya, 2:818 (1962).
24. C. N. Reilley, G. P. Hildebrand, and J. W. Ashley, Anal. Chem., 34:1198 (1962).
25. A. A. Zhukhovitskii and N. M. Turkel'taub, Gas Chromatography [in Russian], Moscow, Gostoptekhizdat (1962).
26. N. M. Turkel'taub, Gas Chromatography, Proceedings of the Second All-Union Conference [in Russian], Moscow, Nauka (1964), p. 88.
27. N. A. Palamarchuk, Gas Chromatography [in Russian], No. 1 (1964), p. 125.
28. M. J. Yanovskii and G. A. Gaziev, Dokl. AN SSSR, 120:812 (1958).
29. M. J. Yanovskii and G. A. Gaziev, Gas Chromatography, Proceedings of the Second All-Union Conference [in Russian], Moscow, Nauka (1964), p. 79.
30. A. A. Zhukhovitskii and N. M. Turkel'taub, Dokl. AN SSSR, 143:646 (1962).
31. A. A. Zhukhovitskii, N. M. Turkel'taub, H. Hauer, M. N. Lagashkina, M. A. Malyasova, and G. P. Shlepuzhnikova, Zav. Lab., 29:8 (1963).
32. A. A. Zhukhovitskii, N. M. Turkel'taub, L. A. Malyasova, M. S. Selenkina, M. M. Lapkin, and A. V. Somov, Gas Chromatography, Proceedings of the Third All-Union Conference [in Russian], Izd. Dzerzh. Fil. OKBA (1966), p. 5.
33. D. C. Myers and F. A. Schmidt-Bleek, Talanta, 13:1695 (1966).
34. V. Willis, Nature, 184:894 (1959).
35. K. Lesser, Angew. Chem., 72:775 (1960).
36. W. S. Galway and J. C. Sternberg, U.S. patent 3,169,832 (1967).
37. H. Gnauk, Gas Chromatography 1961 [collection of Russian translations], Moscow, Gostoptekhizdat (1963), p. 64.
38. M. Hauer, M. N. Lagashkina, B. P. Okhotnikov, and E. P. Fesenko, Gas Chromatography, Proceedings of the Second All-Union Conference [in Russian], Moscow, Nauka (1964), p. 421.
39. J. Klesment, J. Chromat., 31:28 (1967).
40. J. Klesment, J. Chromat., 41:1 (1969).

Chapter III

Gas-Chromatographic Detectors

A gas-chromatographic detector is used to quantitate the results from the separation; it measures the concentration or flux of the compounds after the column.

Two types of differential detector are in use in gas chromatography: (1) those whose response is dependent on the concentration c, and (2) those whose response is dependent on the mass or flux and whose readings are thus dependent on the flow speed [1, 2]. The following are then the relationships for the signal:

$$I_c = K_c c, \tag{36}$$

$$I_I = K_I j = K_f v c, \tag{37}$$

in which K_c and K_f are coefficients of proportionality, v is carrier gas flow rate (by volume), and j is the flow rate of the compound.

Then I_c is only slightly dependent on v, whereas I_f varies rapidly with v. These two types of detector are widely used in various applications, including impurity analysis. Examples are the katharometer and the flame–ionization detector.

The following major characteristics are involved in detector choice:

1. Limit of detection (the c producing a signal twice the fluctuation-noise level);
2. The fluctuation-noise level (the amplitude of the short-period signal variations when the detector receives pure carrier gas);

3. The linear dynamic range (the range in c where the response is proportional to c);
4. The detection time constant (the time to attain 0.698
   of the change in response to an instantaneous change
   in c to a new steady value).

The limit of detection is important in impurity analysis [2a],
but sometimes the limiting factor in a fast analysis is the lag in
response. We can neglect this lag if [3]

$$\tau \geqslant 2.6T, \tag{38}$$

where τ is peak width in sec and T is the deflection time of the recorder carriage in sec.

Zizin [4, 5] has given a detailed analysis of the distortion in
peak shape and position caused by katharometer lag, and his method can be applied to the lag in other types of detector.

Concentration detectors usually have low sensitivity, but they
continue to be widely used in chromatography [6], including impurity analysis.

The Katharometer. This employs measurement of the
electrical resistance of a filament, which is affected by the thermal conductivity of the gas from the column, which itself is dependent on c. A differential system is commonly employed, in which
the working and reference elements are connected in a Wheatstone
bridge, with the reference cell receiving the pure carrier gas. The
bridge is balanced when both cells receive pure carrier. An eluted
compound alters the temperature and resistance of the spiral, and a
recorder registers the out-of-balance signal, which is proportional
to c and is given [7, 8] by

$$E_0 = K (\Delta T)^{3/2} (\lambda_g)^{1/2} \left(\frac{\lambda_g}{\lambda_m} - 1 \right), \tag{39}$$

where K is the cell constant, ΔT is the temperature difference between the wire and the cell wall, λ_g is the thermal conductivity of
the pure gas, and λ_m is the same for the mixture.

The sensitivity is controlled by λ_g/λ_m and is usually best if
λ_g is high (helium or hydrogen) [9-12]. The response is also pro-

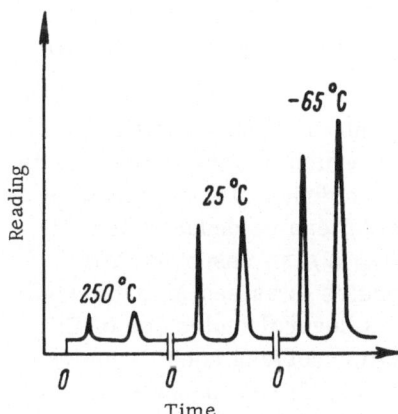

Fig. 13. Change in response caused by detector temperature change [13] for a column containing 5A molecular sieve at 25° operating with helium as carrier and air as sample.

portional to ΔT, so measurements are usually made at fairly high currents I (E_0 is proportional to I^3 [13, 14]). It has recently been shown [13] that the sensitivity to a low-boiling compound can also be raised by cooling the body of the detector. Figure 13 shows curves for separation of the components of air on 5A molecular sieve. Cooling increases the response by a factor six.

It is possible to assay impurities at the $10^{-2}\%$ level when an ordinary type of katharometer is used [15-25]. If special care is taken (thermostatic control, current stabilization, etc.) and the katharometer is of high sensitivity, it is possible to reach $10^{-3}\%$, or even $5 \cdot 10^{-4}\%$ if thermistors are used [26, 27].

The katharometer has the advantage of being universal; it can be used with permanent gases, various inorganic compounds (including corrosive ones such as NO_2, HCl, and compounds containing F), and the vapors of organic compounds. The signal size is dependent on the nature of the compound [14, 28-32] but is only slightly dependent on q [33]. The compounds may react with the hot wire, and the response can be affected by contamination of the surfaces of wire and channel [34-37], so there is a limit to the increase in response from increasing I.

A density detector has certain advantages [37a].

Katharometer sensitivity can be raised by using reactions of the compounds. The eluted compounds pass through a reaction

space before the detector, where they are converted to simpler
compounds, which may substantially alter the response and simplify quantitative evaluation.

It is common to convert all organic compounds to CO_2 or hydrogen when a katharometer is used, which has the following advantages: (1) tedious calibration is avoided, and the contents of
components of the same class (in wt.%) can be deduced directly
from the peak areas given by the CO_2; (2) the response is improved because the number of molecules is raised (one molecule
of an organic compound usually gives several molecules of CO_2
on combustion) and also because better working conditions can be
used (higher I, lower cell temperature); (3) katharometer design
is simplified, and a low-temperature katharometer can be used
with high-boiling compounds (the converter allows the katharometer to be stabilized at room temperature although the column is
at a high temperature). If further examination of the compounds
is required (e.g., via qualitative reactions), the gas flow can be
divided before the converter. The three main conversion methods
for organic compounds yield CO_2, hydrogen, and methane.

Martin and Smart [38] were the first to use combustion of organic compounds to CO_2 before the detector. Conversion methods
have been surveyed [39], and some applications are given in [40-
42].

Green [43] used double conversion of separated fractions to
increase the response. The organic compound is oxidized to
$CO_2 + H_2O$, and then the gas enters a section containing finely
divided reduced iron, where the water is reduced to hydrogen.
The carrier gas was nitrogen, and the hydrogen was detected.
The converter operated at 700-800°C. The gas from the converter
passed through a short sodalime column to remove the CO_2.

Figure 14 gives curves for various conversion methods, which
raise the response and allow one to perform analyses with smaller
samples, which increases column performance. A layer of silver
may be inserted in the reaction section [44] in analysis for compounds containing Cl and Si.

These methods improve the limit of detection by factors of
10-100, whereas ionization methods give an improvement in the
limiting c by factors of 10^3-10^5. The latter are therefore widely

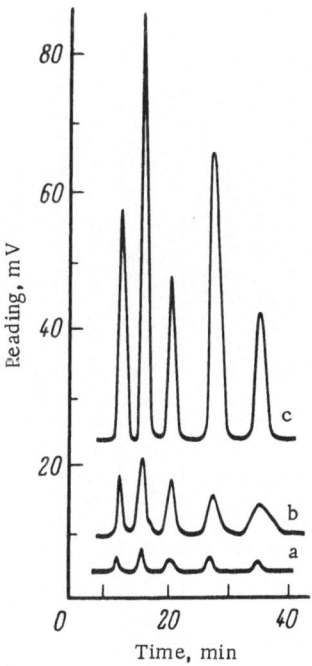

Fig. 14. Differences in response between conversion methods [43]: a) no conversion, b) conversion to CO_2, c) conversion to hydrogen; 1) cinnamaldehyde, 2) butyl benzoate, 3) coumarin, 4) phenylpropyl benzoate, 5) naphthyl phenyl ketone. Column containing 20% silicone oil on celite 545, gas nitrogen, 1 μl sample.

used in analysis for organic impurities, e.g., the flame-ionization and argon detectors. Table 4 gives the basic characteristics of some common detectors, which allow one to evaluate the scope for using each. A high-sensitivity detector gives access to lower limiting c or allows one to use smaller q and thus to perform the separation under more favorable conditions, since the separation performance tends to improve as q is reduced. We therefore consider in more detail the characteristics of some ionization detectors.

The Argon Ionization Detector. Figure 15 shows one form of this. The β particles from a radioactive source ionize and excite argon atoms (the carrier gas). At the same time, argon atoms are excited by collision with electrons that are accelerated by an electric field. Production of the metastable state is basic to a classical argon detector (the ionization energy is 15.7 eV, while the metastable state has an energy of 11.6 eV). The

TABLE 4. Characteristics of Some Common Detectors [11]

Characteristic	Detector type		
	Katharometer	Flame-ionization detector	Electron-capture detector
Detection limit	$1 \cdot 10^{-4}\%$	$1 \cdot 10^{-7}\%$ or 10^{-13} g of carbon per sec	$0.1 \cdot 10^{-7}\%$ (for lindane)
Linear dynamic range	10^4	10^7	~ 10
Time constant	100-250 msec	1 msec (limited by the electrometer)	1-5 sec (limited by detector geometry)
Maximum working temperature, °C	500	500-1000	225°C for detectors with tritium sources, 350°C for nickel sources, 400°C when there is no radioactive source;
Gas flow rate	1 ml/min-1 liter/min	1-200 ml/min	10-200 ml/min
Type of compound	Any	Almost all compounds containing carbon	Pesticides, steroids, nitro compounds, halogen compounds, etc.

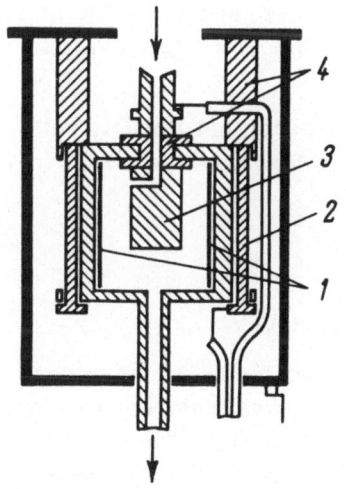

Fig. 15. An argon ionization detector: 1) radioactive foil, 2) brass
chamber, 3) brass electrode, 4)
PTFE parts.

metastable atoms have lifetimes about 10^3 times larger than those
of the other active particles. A metastable atom ionizes an organic molecule by collisions, because most organic compounds have
ionization energies less than 11.6 eV.

The following reactions represent approximately the processes in an argon detector:

1) $Ar + e^{\bullet} \xrightarrow{15.7\,eV} Ar^{+} + 2e$,

2) $Ar + e \xrightarrow{11.6\,eV} Ar^{\bullet} + e$,

3) $Ar^{\bullet} + 2\,Ar \rightarrow 3\,Ar$,

4) $Ar^{\bullet} + $ compound $(M) \rightarrow Ar + M^{+} + e$.

Reactions 1-3 produce the background current, while reaction 4
provides the working signal. See [45-47] for the theory of this
detector, whose signal can [48] be put as

$$I_s = I_b \left[\exp\left(\frac{c}{K_d/K_i + c}\, r_m\right) - 1\right], \tag{40}$$

where I_b is the background current, K_d is the rate constant for
loss of metastable atoms in the body of the detector, K_i is the rate

constant for ionization of organic molecules, and r_m is the mean number of metastable atoms formed by one electron moving in the reaction zone.

It follows from (40) that I_s is, in general, not linearly related to c; but for c small we can put (40) approximately as

$$I_s = I_b \frac{K_i}{K_d} r_m c. \tag{41}$$

The response is thus linear for c small.

The ionization current increases linearly with c up to some limit representing the equilibrium state (equal numbers of argon atoms excited and deexcited in a given interval). The linear dynamic range widens as the electrode voltage increases, while the limit of detection improves. However, the linear dependence on c is lost at high voltages, as I_s increases much more rapidly than c. A continuous discharge sets in at several kV. The following equation represents approximately the sensitivity of an argon detector:

$$S_i = \frac{K}{M_i^{1/6}}, \tag{42}$$

where K is a coefficient of proportionality and M_i is the molecular weight of the compound.

Then (42) gives the variation in S_i between compounds. The relation applies for compounds of almost all classes with M_i up to 100. Deviations occur for nitro-compounds and ones containing halogens, which have high electron affinities. Also, polycyclic aromatic hydrocarbons give the converse dependence of S_i on M_i [49].

A pure carrier gas has to be used with an argon detector. The response is constant if the argon contains less than $3 \cdot 10^{-3}\%$ water, $10^{-2}\%$ oxygen, and $10^{-1}\%$ nitrogen [50]. The response is reduced by a factor 10 when the water content rises from $3 \cdot 10^{-3}\%$ to $10^{-1}\%$. The argon is best dried with a column of 5A molecular sieve.

A classical argon detector has the disadvantage of containing a radioactive source. An interesting design [51] dispenses with that source. The absence of the source plate greatly reduces the background current, increases the stability at high anode voltages, and results in a lower noise level. The response remains at the level found for Lovelock's detector. For instance, $5 \cdot 10^{-4}$ μg of methyl laurate can be detected with an anode potential of 200 V.

The Flame-Ionization Detector. This (Fig. 16) is very widely used [52, 53]. The resistance of a hydrogen flame is greatly reduced by traces of organic compounds that produce ions by combustion [54, 55]. These ions are collected by electrodes, one of which is usually the burner nozzle. A pure hydrogen flame usually gives a background current of the order of 10^{-11} A, the exact value being very much dependent on the purity of the hydrogen, air, and carrier gas. The background current usually fluctuates by not more than $5 \cdot 10^{-13}$ A in response to flow-speed fluctuations. Figure 16 shows the design of the DIP-2.

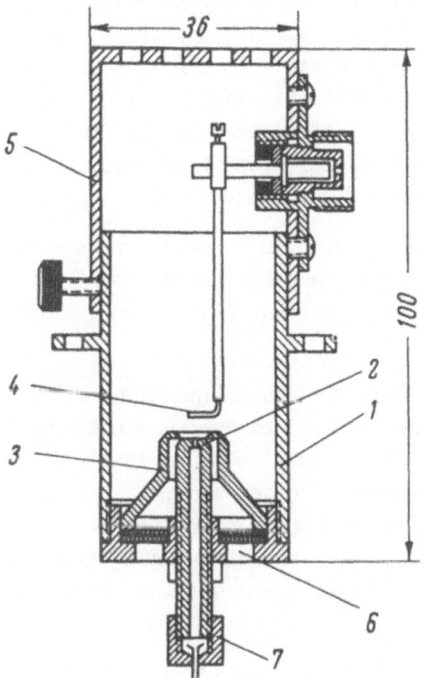

Fig. 16. The DIP-2 flame-ionization detector: 1) body, 2) burner, 3) diffuser, 4) collector, 5) upper removable cover, 6) air inlet, 7) gas inlet.

The linear dynamic range usually attains 10^6 and is largely governed by the geometry of the nozzle [56] and the design of the collecting electrodes [52, 57, 58]. An interesting design of detector has been described [58a].

It is very important to choose the correct speeds for the carrier gas, hydrogen, and oxidant (usually air). It has been shown [59, 60] that the response increases rapidly with the carrier gas speed, and there is a speed giving the highest response. The detector signal increases also with the hydrogen flow [61] up to 30-40 ml/min (the compound determines the exact value). Higher hydrogen speeds have little effect on the response, which is particularly important for precision quantitative work. There is no point in using a rate greater than 77 ml/min, as the background current increases rapidly.

The oxidant purity and flow rate affect the response considerably. The largest response is usually obtained with a 5- to 10-fold excess of air (an even larger excess may be required [62, 63] to produce a signal independent of air flow speed). Careful purification from hydrocarbon contaminants reduces the noise level (the sensitivity is largely unaffected by water vapor [64]); the usual silica gel and molecular sieves do not remove hydrocarbons completely. It is much more effective to pass the gas through an alumina column bearing 15-25% silver nitrate heated to 150-200°C [65]. This produces irreversible chemisorption of organic vapors, which reduces the background current by a factor 8-10.

The flame temperature affects the ionization performance and thus the sensitivity [66], so oxygen may be added to the gas flow before combustion [67], or the air may be completely replaced by oxygen [61, 68].

The response varies with the nature of the compound. The signal is roughly proportional to the number of carbon atoms in the compound (Table 5) [69, 70]. The ionization effect is dependent on the bond type and on the content of other atoms in the molecule; e.g., nitrates give rather lower molar signals than do nitrites and nitroalkanes, and an NO_2 group in the molecule reduces the sensitivity considerably relative to the corresponding hydrocarbon.

The common ionization detectors thus allow one to detect trace impurities. Difficulties arise at present in developing new

TABLE 5. Relative Sensitivity of a Flame-Ionization Detector

Compound	Relative molar sensitivity	Number of C atoms	Compound	Relative molar sensitivity	Number of C atoms
Methane	1.00	1	Carbon monoxide	0.0	1
Ethane	2.05	2	Carbon dioxide	0.0	1
Ethylene	2.05	2	Nitric oxide	0.0	0
Acetylene	2.3	2	Carbon disulfide	0.0	1
Propane	3.15	3	Nitromethane	0.49	1
Propylene	3.05	3	Nitroethane	1.4	2
Cyclopropane	3.25	3	1-Nitropropane	2.3	3
Butane	4.2	4	2-Nitropropane	1.95	3
Isobutane	4.2	4	1-Nitrobutane	2.95	4
But-1-ene	4.2	4	2-Nitrobutane	3.2	4
But-2-ene	4.1	4	Methyl nitrate	0.36	1
Isobutylene	4.1	4	Ethyl nitrate	1.2	2
But-1,3-diene	4.2	4	1-Propyl nitrate	1.95	3
Hexane	6.6	6	2-Propyl nitrate	2.3	3
Benzene	6.1	6	1-Butyl nitrate	2.3	4
Cyclohexane	6.6	6	1-Propyl nitrite	2.3	3
Heptane	7.8	7	2-Propyl nitrite	2.2	3
Methanol	0.87	1	2-Butyl nitrate	3.5	4
Ethanol	2.05	2	Isobutyl nitrite	2.9	4
Carbon tetrachloride	0.67	1	t-Butyl nitrite	3.75	4
Chloroform	0.89	1	Isobutyraldehyde	2.8	4
Freon-12	0.42	1	t-Butanol	3.95	4
Vinyl chloride	1.85	2	Isobutylene oxide	2.85	4

methods mainly not on account of inadequate detector response but because the impurity is overlapped by the major component or the impurities are inadequately separated, i.e., because the sorbents and detectors are of inadequate selectivity.

LITERATURE CITED

1. J. Halasz, Anal. Chem., 36:1428 (1964).
2. V. I. Kalmanovskii, Trudy po Khim. i Khim. Tekhnol. (Gor'kii), 3:3 (1960).
2a. A. J. P. Martin, in: Gas Chromatography in Biology and Medicine, edited by R. Porter, London, J. and A. Churchill Ltd. (1969), p. 2.
3. I. S. Abrosimov and L. N. Mogilevskii, Izv. AN SSSR, Ser. Fiz., 19:49 (1955).
4. V. G. Zizin, Ph.D. Thesis, Moscow, All-Union Nuclear Geology and Geophysics Research Institute (1965).

5. V. G. Zizin, "Processing products from sulfur-bearing oils," Trudy BashNII Neft. Prom., No. 6, Moscow, Gostoptekhizdat (1963), p. 152.
6. The Hot Wire (Gow-Mac Instrument Company), 1:2 (1965).
7. C. Bokhoven and A. Dijkstra, Nature, 186:793 (1960).
8. L. I. Schmauch and K. A. Dinerstein, Anal. Chem., 32:343 (1960).
9. G. Dijkstra, Vapour Phase Chromatography, 1956, edited by D. H. Desty, New York, Academic Press (1957), p. 74.
10. N. H. Ray, Nature, 182:1663 (1958).
11. N. H. Ray, Nature, 183:674 (1959).
12. W. A. Wiseman, Nature, 183:1321 (1959).
13. A. E. Lawson and J. M. Miller, J. Gas Chromat., 4:273 (1966).
14. L. C. Browning and J. O. Watts, Anal. Chem., 29:24 (1957).
15. A. V. Alekseeva and K. A. Gol'bert, Zav. Lab., 27:972 (1961).
16. G. Nodop, Z. Anal. Chem., 164:120 (1958).
17. C. E. Bennet, S. D. Nogare, L. W. Safranski, and C. D. Lewis, Anal. Chem., 30:898 (1958).
18. L. M. Kontorovich and A. V. Iogansen, Zav. Lab., 28:146 (1962).
19. F. A. Fabrizo, K. W. King, C. C. Cerato, and J. W. Loveland, Anal. Chem., 31:2060 (1959).
20. H. K. Kaufman and A. Zlatkis, Chem. Ind. (London), 1958:1001.
21. D. M. G. Lawrey and C. C. Cerato, Anal. Chem., 31:1011 (1959).
22. P. I. Markosov, V. N. Zaichenko, and V. A. Lityaeva, Zav. Lab., 27:285 (1961).
23. J. A. Schlos, Anal. Chem., 33:359 (1961).
24. D. J. Timms, J. H. Konrath, and C. K. Chirnside, Analyst, 83:600 (1958).
25. M. Tath and C. Mever, Erdöl Z., 76:37 (1960).
26. E. Bia, P. Manaresi, and L. Motta, Anal. Chem., 31:1910 (1959).
27. W. F. Wilhite, J. Gas Chromat., 4:47 (1966).
28. E. G. Hoffmann, Anal. Chem., 34:1216 (1962).
29. A. E. Messner, D. D. Rosie, and P. A. Argabright, Anal. Chem., 31:230 (1959).
30. K. L. Grob, D. Mercer, T. Gribbon, and H. Wells, J. Chromat., 3:545 (1960).
31. H. Veening and G. D. Dupre, J. Gas Chromat., 3:153 (1966).
32. L. D. Hinshaw, J. Gas Chromat., 4:300 (1966).
33. B. D. Smith and W. W. Bowden, Anal. Chem., 36:87 (1964).
34. J. G. Keppler, G. Dijkstra, and J. A. Schols, Vapour Phase Chromatography, 1956, edited by D. H. Desty, New York, Academic Press (1957), p. 222.
35. J. S. Parsons, W. B. Prescott, and H. C. Lawrence, Anal. Chem., 34:1337 (1962).
36. K. G. Ackman, R. D. Burgher, J. C. Sipos, and P. H. Odense, J. Chromat., 9:531 (1962).
37. R. G. Ackman and R. D. Burgher, Anal. Chem., 35:413 (1963).
37a. A. J. P. Martin and A. T. James, Biochem. J. (London), 63:138 (1956).
38. A. E. Martin and J. Smart, Nature, 175:422 (1955).
39. M. C. Simmon, L. M. Taylor, and N. Narer, Anal. Chem., 32:731 (1960).
40. J. R. Hunter, V. H. Ortegren, and T. W. Pence, Anal. Chem., 32:682 (1960).
41. W. Stuve, Gas Chromatography: Papers at the Second International Symposium in Amsterdam and the Conference on the Analysis of Mixtures of Volatile Substances in New York [Russian translation], Moscow, IL (1961), p. 168.

42. S. I. Krichmar and M. I. Beilina, Zav. Lab., 26:1171 (1960).

43. J. E. Green, Nature, 175:295 (1957).

44. J. Franz and M. Wurst, Gas Chromatography, Proceedings of the First All-Union Conference [in Russian], Moscow, Izd. AN SSSR [1960], p. 289.

45. J. E. Lovelock, A. T. James, and E. A. Piper, Gas Chromatography [Russian translation], Moscow, IL (1961), p. 406.

46. J. E. Lovelock, Gas Chromatography: Proceedings of the Third International Symposium [Russian translation], Moscow, Mir (1964), p. 27.

47. Z. I. Knapp and A. S. Meyer, Anal. Chem., 36:1430 (1964).

48. V. R. Rotin, "Automation in technological processes," Trudy VNIIKANeftegaz, No. 2, Moscow, Nedra (1968), p. 201.

49. J. E. Lovelock, A. T. James, and E. A. Piper, Ann. N. Y. Acad. Sci., 72:720 (1959).

50. R. S. Evans and P. G. W. Scott, Nature, 190:710 (1961).

51. E. Haahti, T. Nikkari, and E. Kulonen, J. Chromat., 3:372 (1960).

52. J. Harley, W. Nel, and V. Pretorius, Nature, 181:117 (1958).

53. J. G. McWilliam and K. A. Dewar, Nature, 182:1664 (1958).

54. G. Schemburg, Z. Anal. Chem., 189:14 (1962).

55. J. C. Sternberg, W. S. Gallaway, and D. T. L. Jones, ISA Proceedings of the Third International Gas Chromatography Symposium 1961, Pittsburgh (1961), p. 159.

56. J. Novák and J. Janák, Chem. Listy, 57:371 (1963).

57. A. E. Thompson, J. Chromat., 2:148 (1959).

58. J. Middlehurst and B. Kennet, J. Chromat., 10:294 (1963).

58a. J. Middlehurst and B. Kennet, J. Chromat., 10:294 (1963).

59. J. G. McWilliam and K. A. Dewar, Gas Chromatography, Proceedings of 2nd Symposium Amsterdam, edited by D. H. Desty, London, Butterworths (1958), p. 146.

60. V. I. Kalmanovskii, Gas Chromatography, Proceedings of the Second All-Union Conference [in Russian], Moscow Nauka (1964), p. 410.

61. H. Bruderreck, W. Schneider, and J. Halasz, Anal. Chem., 36:461 (1964).

62. O. Hainová, P. Bocek, J. Novák, and J. Janák, J. Gas Chromat., 5:401 (1967).

63. Z. Batt and K. R. Cruickhank, J. Chromat., 21:296 (1916).

64. D. W. Hill and H. A. Newell, Nature, 206:708 (1965).

65. R. L. Hoffmann, J. R. List, and C. D. Evans, J. Gas Chromat., 5:383 (1967).

66. G. C. Sternberg, W. S. Gallaway, and D. T. L. Jones, Gas Chromatography, New York, Academic Press (1962), Chapter 18.

67. R. L. Hoffmann and C. D. Evans, J. Gas Chromat., 4:318 (1966).

68. K. Jones and R. Green, Nature, 210:1355 (1966).

69. A. J. Andreatch and R. J. Cvetanovic, J. Gas Chromat., 32:1021 (1960).

70. G. Paraskevopoulos and R. J. Cvetanovic, J. Gas Chromat., 25:479 (1966).

71. T. Johns and J. C. Sternberg, Instrumentation in Gas Chromatography, Eindhoven, Centrex Publishing Company (1967), p. 179.

Chapter IV

Use of Selective Sorbents
and Selective Detectors

There are detailed surveys [1-5] on the best sorbents for separating substances present in relatively high concentrations.

New and specific problems arise in analyzing specially pure materials, quite apart from the usual problems of separating numerous substances with similar properties; these occur because the broad band of the main component tends to mask the others. The sorbent therefore has to be especially selective. Sorbents may be divided into those where the relevant impurities run ahead of the main component and those where they run behind the latter. There is no particular difficulty from masking in impurity analysis in the first case, so the static liquid phase is often chosen on that basis. For instance, traces of acetylene in ethylene have [6] been assayed with 30% heptadecane on Inza earth, on which the acetylene runs ahead of the ethylene, and a reliable quantitative assay can be performed.

Figure 17 [17] shows curves for trace impurities in toluene separated with (a) an unselective liquid and (b) a selective one; 10% benzyldiphenyl on celite 545 selectively retains aromatic compounds and allows about 15 impurities in toluene to be assayed. This example illustrates clearly the importance of selectivity in impurity analysis.

Unfortunately, it is far from always possible to choose a selective sorbent giving the impurities first, especially if these are

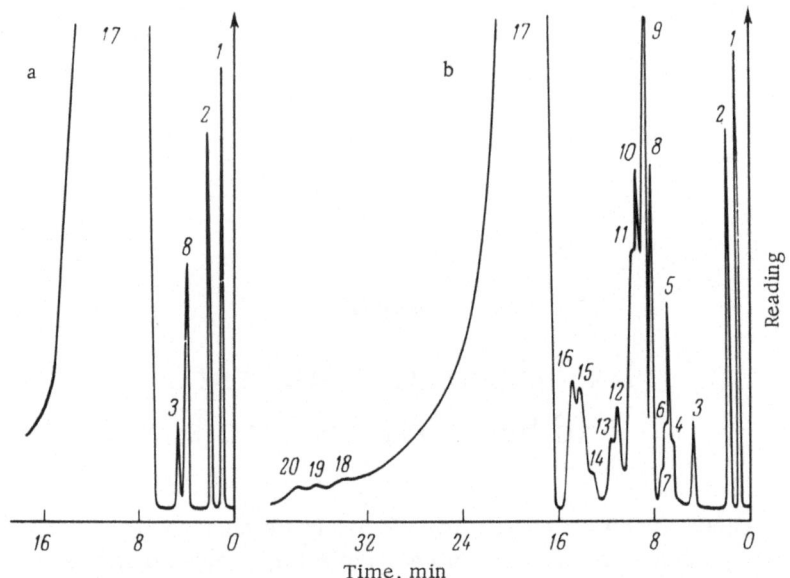

Fig. 17. Assay of impurities in toluene: a) 10% Apiezon L on celite 545, 2.7 m
column at 100°C; b) 10% benzyldiphenyl on celite 545, 2.7 m column at 75°C;
1) methane, 2) n-pentane, 3) heptane (internal standard), 4) methylcyclohexane,
5) 2-methylheptane + 3-methylheptane, 6) 4-methylheptane, 7) ethylcyclopentane, 8) benzene, 9) 1,2,4-trimethylcyclopentane, 10) n-nonane, 11) 1,1-di-
methylcyclohexane, 12) trans-1,2-dimethylcyclohexane, 13) trans-1,3-dimethyl-
cyclohexane, 14) cis-1,2-dimethylcyclohexane, 15) n-propylcyclopentane, 16)
ethylcyclohexane, 17) toluene, 18) cyclooctane, 19) 1,3-dimethylbenzene, 20)
1,1,3-trimethylcyclohexane.

similar in properties to the base. In that case, one tries to choose
a system where the impurities run well behind the base.

Here a measure of the selectivity is

$$\theta = t_i/t_0, \tag{43}$$

where t_i is the retention time for the first relevant impurity after
the base and t_0 is the retention time for the base. The separation
condition is

$$\mu_b < t_i - t_0, \tag{44}$$

where μ_b is the base band width, i.e., the region on the chromato-gram where impurities cannot be assayed on account of overlap. This width (in time terms) must be less than the difference be-tween the retention times. It follows from (43) and (44) that

$$\theta > 1 + \frac{\mu_b}{t_0}. \tag{45}$$

A system with θ large is to be preferred. If the column perfor-mance for the base increases (μ_b decreases), less selectivity is needed, so particular attention should be given to reducing μ_b.

Published tables of relative retention times should be used in the preliminary choice of the liquid phase, as selective liquids of various types are available (coordinating compounds, liquid crys-tals, volatile solvents, etc.); a good survey [8] has been published. Superselective liquids are provided by: (1) solutions of silver salts in benzyl cyanide, which separate olefins well and hardly hold back saturated hydrocarbons, (2) tetracyanoethylene pentaerythritol, which has alkanes, olefins, and cycloalkanes running ahead of ben-zene. The selectivity of solid sorbents has also been discussed [5].

Sometimes the choice of liquids or adsorbents is so restricted that one cannot devise a solution on the basis of a single column, in which case a multistage system may be used [9]. Here major contributions have been made by Vigdergauz et al. [3, 10-12].

The final result from a chromatographic analysis is deter-mined by the characteristics of the sorbent and the detector. Im-portant impurities should be assayed by using a selective sorbent having marked differences between the retention times for the im-purities and base. Another approach is to use a selective detector whose sensitivity to an impurity is much higher than that to the base. In that case the impurity can be determined even if it is overlapped by the base.

The following are relevant details of the selective response of the detectors most commonly used in gas chromatography.

The katharometer is virtually unresponsive to any com-pound whose thermal conductivity is close to that of the carrier gas. This makes it desirable, for instance, to use pure ethylene as the carrier in determining traces of hydrogen in ethylene; sim-ilarly, pure helium is used to determine contaminants in helium.

The flame-ionization detector is very sensitive to organic compounds, but it cannot be used to record many common substances: nitrogen, helium, hydrogen, oxygen, inert gases, water, freon, H_2S, $SiCl_4$, trichlorosilane, SiF_4, COS, phosgene, formic acid, SO_3, CO, and CO_2 [13].

Recently, much use has been made of selective detectors having high sensitivity only to certain groups of compounds [13a, 13b]. Cremer [14] considered that a detector determines compound i selectively with respect to compound j if

$$S = \left(\frac{S_i}{q_i}\right) \Big/ \left(\frac{S_j}{q_j}\right) > 10 , \qquad (46)$$

where S_i and S_j are the areas of the peaks for i and j with sample sizes q_i and q_j respectively.

The following are the selective detectors most widely used in gas chromatography.

The electron-capture detector. This was first described by Lovelock and Lipsky [15], and it employs the marked reduction in the current in an ionization chamber when any substance of high electron affinity is present [15-18].

The β particles from a radioactive source ionize the gas and produce thermal electrons, which move to the anode in response to the electric field. If a compound having a high electron affinity enters, negative ions are formed by electron capture, and so the concentration of free thermal electrons is reduced. Although a molecule may have only a small probability of trapping a free electron, the electron-molecule collisional frequency is very high at atmospheric pressure, so there is nearly 100% probability of producing negative molecular ions for substances of high electron affinity.

The reduced electron concentration results in a reduced ionization current in accordance with

$$I = I_0 \exp\left(- Ec\, \chi\right), \qquad (47)$$

where I_0 is the saturation current for the pure carrier gas, I is the current in the presence of a compound at a concentration c, χ is a

parameter dependent on the chamber geometry, and E is the electron absorption coefficient, which is dependent on the working voltage and on the electron affinity.

The detector is usually operated at 10-100 V. Recently, pulsed
operation has been used (0.5 μsec pulses separated by 100 μsec
[19]). The gas is argon mixed with a compound that does not trap
electrons, e.g., 5-10% methane.

There is a detailed literature survey on this detector [20].
The sensitivity is higher than that of a flame-ionization detector,
e.g., the minimum detectable flow rate for CCl_4 is $3 \cdot 10^{-14}$ g/sec.
However, the linear dynamic range is not more than 10^3, as against
10^6 for flame ionization.

The electron affinity is very much dependent on the nature and
position of any substituents in the molecule. The sensitivity falls

TABLE 6. Relative Response of an Electron-Capture Detector
to Various Compounds

Compound	Relative molar response	Compound	Relative molar response
1-Chlorobutane	1.0	Bromocyclopentane	280
2-Chlorobutane	2.0	1-Bromoprop-2-ene	$4 \cdot 10^3$
1-Chloro-2-methylpropane	1.7	1,1-Dibromoethane	$1.1 \cdot 10^5$
2-Chloro-2-methylpropane	12	1-Iodobutane	$9 \cdot 10^4$
1-Chloropentane	1.0	Benzene	0.06
1-Chlorohexane	1.1	Toluene	0.2
1-Chloroheptane	1.5	2-Fluorotoluene	0.55
1-Chlorooctane	1.6	4-Fluorotoluene	0.55
1,2-Dichloroethane	190	Chlorobenzene	75
1,4-Dichlorobutane	15	Bromobenzene	450
1,1-Dichlorobutane	110	1-Butanol	1.0
trans-1,2-Dichloroethylene	370	Di-n-butyl ether	0.6
cis-1,2-Dichloroethylene	90	Acetone	0.5
Chloroform	$6 \cdot 10^4$	Methyl butyrate	0.9
Carbon tetrachloride	$4 \cdot 10^5$	2,3-Butadiene	$5 \cdot 10^4$
1-Bromopropane	255	n-Propyl pentafluoropropionate	450
1-Bromobutane	280	n-Heptyltrifluoroacetone	4.5

rapidly in the sequence I > Br > Cl > F and increases with the number of halogen atoms in the molecule [21]. The sensitivity to iodine compounds is 10^7 times that for fluorine ones [22]. Some compounds have little or no electron affinity (Table 6 [23]), especially hydrocarbons, alcohols, esters, amines, and thioesters. The response to these can be raised by converting them to compounds with high electron affinity, e.g., halogen derivatives. In this way it was possible [24] to produce a 1000-fold improvement in the response to styrene (relative to a flame-ionization detector).

The high sensitivity and selectivity have given the detector many uses in impurity analysis, e.g., for toxic compounds [25-30], organometallic compounds [24, 31-35], and in biochemistry and medicine [36-39]. However, the main use of this detector is in analysis of food, blood, etc., for various pesticides [40-48].

The sodium thermionic detector. This employs the selective increase in the current in a flame-ionization detector having one electrode coated with a salt of an alkali or alkaline-earth metal; it is used in the analysis of compounds containing phosphorus or halogens [49, 50].

Two types of this detector have been described: (1) with a single nozzle, above which is an electrode treated with the alkali-metal salt, (2) with two nozzles one above the other and an upper grid of platinum or stainless steel treated with the salt.

The first type has the advantage that it is easily made up from an ordinary flame-ionization detector. The modification increases the response by a factor 600 for phosphorus compounds with one P atom per 10 carbon atoms and by about a factor 20 for compounds containing six chlorine atoms. Hydrocarbons produce virtually no response.

The second type has two independent systems, so it can be used in ordinary analyses and also in selective analyses for compounds containing phosphorus or halogen. This detector is 100-1000 times more selective for such compounds than a one-flame detector.

This detector has been used in selective assay of pesticides [51, 52].

It has been shown [53] that salts of other alkali metals allow one to increase selectively the response to nitrogen compounds.

The best results are obtained with rubidium and potassium sulfates. Rubidium sulfate on the nozzle raises the response to some nitro compounds by 2-3 orders of magnitude relative to an ordinary flame-ionization detector.

Mass spectrometer [13b]. A promising trend in detector design is the use of devices that not only have high selectivity and sensitivity but also provide information on the nature of the compound.

A double-beam mass spectrometer is very sensitive, and its selectivity can be varied within wide limits. It can provide identification of impurities. Tal'roze et al. [54-57] have developed this approach. Figure 18 [56] shows curves recorded with a Khromass-22. A mass spectrometer allows one to detect impurities overlapped by the base, while peaks can be interpreted from the intensity ratios of the mass lines.

It is possible to assay $10^{-4}\%$ of an impurity with a mass spectrometer operating on single lines with measurement of the ion current by an electrometer amplifier, or $10^{-6}\%$ if a secondary-electron multiplier is used [58].

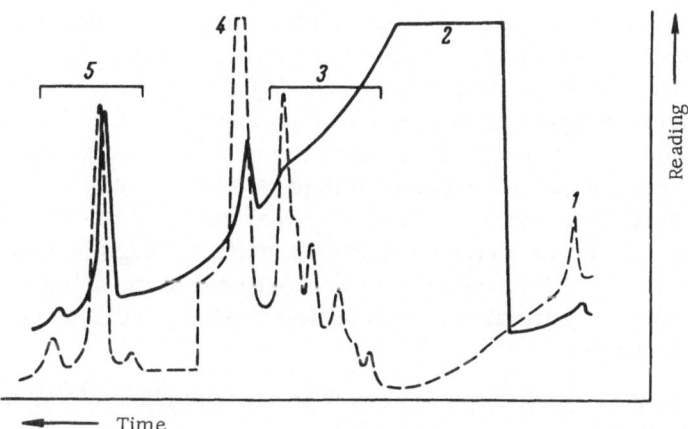

Fig. 18. Double chromatogram from purified benzene as recorded via the lines for 39 mass units (solid curve) and 41 units (broken curve) [56]: 1) n-hexane, 2) benzene, 3) isomeric heptanes, 4) heptane, 5) C_7-C_8 cycloalkanes. Capillary column (30 m) containing squalane at 70°C, 50-μl sample.

Flame-photometer detector. This employs the emission produced from a hydrogen—air flame by compounds containing P and S [59, 60].

There is hardly any response to C, H, N, O, and Cl, so the detector can be used to assay compounds of P and S containing these other elements. It can detect $1\text{-}10 \cdot 10^{-4}\%$ of sulfur compounds and $0.1\text{-}1 \cdot 10^{-4}\%$ of phosphorus ones.

The column gas is first mixed with oxygen (or air) and is burned in a hydrogen flame. The emission is passed through a filter (526 mμ for P compounds, 394 mμ for S compounds) to a photomultiplier and thence to a recorder. The peak area is dependent on the amount of P or S in the band.

A collector electrode can be used with the photomultiplier to use the flame simultaneously in the ionization and photometric modes; also, two filters and multipliers can be used to record P and S impurities present together.

The microcoulometric detector allows one to detect selectively compounds whose molecules contain Cl, S, N, or P [61-65]. The sensitivity is high, so impurities at the 10^{-3} to $10^{-4}\%$ level can be assayed.

The gas is mixed with a gaseous reagent (oxygen or hydrogen) and then passes to a heated reaction tube, where the compounds are converted to simpler forms, e.g., sulfur compounds to SO_2 by oxidation, nitrogen, compounds to NH_3 by reduction, which are then selectively detected by absorption in an appropriate reagent solution. The resulting change is recorded by a pair of indicator electrodes, and the out-of-balance voltage is used to apply a compensating voltage to a pair of electrodes that generate the reagent (restore the initial level). The compensating voltage is also passed to a recorder. The peak area is proportional to the amount of electricity needed to titrate the conversion product, and hence to the element content.

Oxidative conversion is used with various electrolytes to determine S and halogens, while reductive conversion is used with P and N. Phosphorus (as phosphine) is assayed selectively by selective absorbers: HCl is held back by silica gel, and H_2S by alumina or ascarite.

The detector has further advantages in that it allows relatively volatile liquids to be used (these are often more selective). Also, the readings are independent of column temperature and gas speed.

Other detectors. An electron-mobility detector [66-68] has been used to detect traces of water, CO_2, and oxides of nitrogen, which produce virtually no response with ionization detectors. The device employs the reduction in an ionization current, and the limit of detection for CO_2 may be 10^{-9} g/sec in the indirect technique, or 10^{-11} g/sec in direct detection [17].

A low detection limit (around $3 \cdot 10^{-9}$ g/sec) and a wide linear dynamic range (10^3-10^4) occur in the photoionization detector [69-71], which employs ionization by photons of appropriate energy. The detector is not applicable to permanent gases, water vapor, and other compounds with high ionization potentials.

A detector can be based on the effects of compounds on discharge characteristics [72-76]. The discharge detector responds to 10^{-10} to 10^{-12} mole/sec of organic and inorganic compounds [77], and the response is almost temperature-independent over the range 20-600°C [78]; but the device has not been widely used because the stability is poor.

Detectors selective for halogen compounds have been described [79, 80], which respond to about 10^{-7} mg/ml of CCl_4, $SiCl_4$, $GeCl_4$, and $SnCl_4$, while the linear dynamic range is not less than 10^4.

Electrochemical detectors are also of considerable interest. Detectors have been described for selective assay of S and halogens [79, 80] and alcohols and aldehydes [14].

In difficult cases, it is best to use several selective detectors of different types connected in series, which provides a good solution to qualitative and quantitative assay of impurities.

LITERATURE CITED

1. S. D. Nogare and R. S. Juvet, Gas-Liquid Chromatography, Theory and Practice (1962).
2. A. A. Zhukhovitskii and N. M. Turkel'taub, Gas Chromatography [in Russian], Moscow, Gostoptekhizdat (1962).

3. K. A. Gol'bert and M. S. Vigdergauz, Textbook of Gas Chromatography [in Russian], Moscow, Khimiya (1967).

4. E. Leibnitz and H. G. Struppe, Handbuch der Gas-Chromatographie, Leipzig, Akademische Verlagsgesellschaft (1966).

5. A. V. Kiselev and Ya. I. Yashin, Gas-Adsorption Chromatography [in Russian], Moscow, Nauka (1967).

6. M. S. Vigdergauz and L. V. Andreev, Zav. Lab., 31:550 (1965).

7. V. G. Berezkin, B. V. Strizhkov, and V. S. Tatarinskii, Neftepererabotka i Neftekhimiya, No. 8, 26 (1967).

8. B. L. Karger, Anal. Chem., 39:24A (1967).

9. G. Heuschkel, J. Wolny, and S. Skoerowski, Erdöl u. Kohle, 13:98 (1960).

10. M. S. Vigdergauz, K. A. Gol'bert, I. M. Savina, M. I. Afanas'ev, and R. A. Zimin, Zav. Lab., 28:149 (1962).

11. M. S. Vigdergauz and K. A. Gol'bert, Neftekhimiya, 1:706 (1961).

12. M. S. Vigdergauz and M. I. Afanas'ev, Zh. Anal. Khim., 19:1122 (1964).

13. R. D. Condon, P. R. Scholly, and W. Averill, Gas Chromatography, 1960 edited by R. P. W. Scott, London, Butterworths (1960), p. 30.

13a. M. Keejci and M. Dressler, Chromat. Rev., 13:1 (1970).

13b. L. S. Ettre and W. H. McFadden, Ancillary Techniques Gas Chromatography, New York, Wiley (1970).

14. E. Cremer, J. Gas Chromat., 5:329 (1967).

15. J. E. Lovelock and S. R. Lipsky, J. Amer. Chem. Soc., 82:431 (1960).

16. J. E. Lovelock, Nature, 189:729 (1961).

17. J. E. Lovelock, Anal. Chem., 33:162 (1961).

18. N. L. Gregory and J. E. Lovelock, Anal. Chem., 33:45A (1961).

19. J. E. Lovelock, Anal. Chem., 35:474 (1963).

20. V. V. Brazhnikov and K. I. Sakodynskii, Gas Chromatography [in Russian], No. 6, Moscow, NIITÉKhim (1967), p. 11.

21. D. L. Petitjen and C. D. Lants, J. Gas Chromat., 1:23 (1963).

22. J. A. Le Silva, M. A. Schwartz, V. Stefanovic, W. Kaplan, and L. D. Aronte, Anal. Chem., 36:2099 (1964).

23. J. W. Ralls and A. J. Cortes, J. Gas Chromat., 2:132 (1964).

24. H. J. Dawson, Anal. Chem., 36:1852 (1964).

25. J. E. Lovelock and N. L. Gregory, Gas Chromatography, New York, Academic Press (1962), p. 219.

26. C. A. Clemons and A. P. Altschuller, Anal. Chem., 38:133 (1966).

27. P. Devaux and G. Guichon, J. Gas Chromat., 5:341 (1967).

28. K. D. Stewart, J. D. Swank, C. B. Robuts, and H. C. Dodd, Nature, 198:696 (1963).

29. L. J. Priestly, F. E. Oritchfild, and N. H. Ketcham, Anal. Chem., 37:70 (1965).

30. B. J. Gudzinowicz and S. J. Clark, J. Gas Chromat., 3:147 (1965).

31. E. F. Darby, K. A. Kettner, and E. R. Stephens, Anal. Chem., 35:589 (1963).

32. D. S. Abbot, R. B. Harrison, T. O. Totton, and J. Thomson, Nature, 208:13A (1965).

33. E. A. Bacttner and F. C. Dallas, J. Gas Chromat., 3:190 (1965).

34. R. Mochier and R. Sievers, Gas Chromatography of Metal Chelates [Russian translation], Moscow, Mir (1967).

35. E. J. Bonelli and N. Hartmann, Anal. Chem., 35:1980 (1963).
36. H. J. Dawson, Anal. Chem., 35:542 (1963).
37. J. E. Lovelock and A. Zlatkis, Anal. Chem., 33:1958 (1961).
38. W. D. Ross, Anal. Chem., 35:1596 (1963).
39. J. E. Lovelock, P. G. Simmonds, and W. J. A. Vanden Heuvel, Nature, 197:249 (1963).
40. S. J. Clark and H. H. Wotiz, Steroids, 2:535 (1963).
41. M. C. Bowman and M. J. Beroza, J. Assoc. Offic. Agr. Chemists, 48:922 (1965).
42. D. D. Clarke, S. Wilk, and S. E. Gitlow, J. Gas Chromat., 4:310 (1966).
43. H. P. Burchfield and Eleanor E. Storrs, Biochemical Applications of Gas Chromatography, New York, Academic Press (1962).
44. J. E. Barney, C. W. Stanley, and C. E. Cook, Anal. Chem., 35:2206 (1963).
45. H. P. Burchfield, D. E. Johnson, R. J. Wheeder, and J. W. Rhoades, Anal. Chem., 38:28 (1965).
46. B. E. Langlois, A. R. Stemp, and B. J. Liska, J. Milk Food Technol., 27:202 (1964).
47. G. L. Resnick, D. Corbin, and D. H. Sandberg, Anal. Chem., 38:582 (1966).
48. E. S. Goodwik, R. Goulden, and J. Reynolds, Analyst, 86:697 (1961).
49. L. Giuffrida, J. Assoc. Offic. Agr. Chemists, 50:293 (1967).
50. A. Karmen, Anal. Chem., 36:1416 (1964).
51. C. H. Hartman, Bull. Environ. Contamination Toxicol., 1:4 (1966).
52. E. Cremer and H. L. Gruber, J. Gas Chromat., 2:8 (1965).
53. W. A. Aue, C. W. Gehrke, R. C. Tindle, D. L. Stalling, and C. D. Ruyle, J. Gas Chromat., 5:381 (1967).
54. V. L. Tal'roze, V. V. Reznikov, and G. D. Tantsyrev, Dokl. AN SSSR, 159:182 (1964).
55. V. L. Tal'roze, G. D. Tantsyrev, and V. I. Goshkov, Zh. Anal. Khim., 20:103 (1955).
56. G. D. Tantsyrev, V. I. Goshkov, S. T. Kozlov, and V. L. Tal'roze, Zh. Anal. Khim., 21:989 (1966).
57. G. D. Tantsyrev, V. I. Goshkov, S. T. Kozlov, and V. L. Tal'roze, Zh. Anal. Khim., 21:1113 (1966).
58. V. I. Goshkov and S. T. Kozlov, Abstracts of Papers for the International Symposium on Chromatographic Mass Spectrometry [Russian translation], Moscow, Nauka (1968), p. 14.
59. S. S. Brody and J. E. Chaney, J. Gas Chromat., 4:42 (1966).
60. M. C. Bowman and M. Beroza, J. Assoc. Offic. Analyt. Chemists, 50:1228 (1967).
61. O. E. Piringer, E. Tataru, and M. Pascalan, J. Chromat., 2:104 (1964).
62. H. P. Burchfield and R. J. Wheeler, J. Assoc. Offic. Analyt. Chem., 49:651 (1966).
63. R. L. Martin, Anal. Chem., 38:1209 (1966).
64. D. M. Coulson, I. A. Cavanagh, J. E. de Vries, and G. Wilther, J. Agr. Food Chem., 8:399 (1960).
65. S. I. Krichmar and V. E. Stepanenko, Authors' certificate 177,152 (1964); Byull. Izobr., No. 24, 99 (1965).
66. V. Willis, Nature, 183:1754 (1959).
67. J. E. Lovelock, Nature, 187:49 (1960).

68.   V. Yu. Orlov, R. A. Ivanova, and V. V. Brazhnikov, Gas Chromatography,
      Proceedings of the Third All-Union Conference [in Russian], Izd. Dzerzh. Fil.
      OKBA (1966), p. 371.
69.   J. E. Lovelock, Nature, 183:1754 (1959).
70.   V. E. Kazakevich, Gas Chromatography [in Russian], No. 2, Moscow, NIITÉKhim
      (1964), p. 47.
71.   M. Jamane, J. Chromat., 14:355 (1964).
72.   J. Harley and V. Pretorius, Nature, 178:1244 (1956).
73.   R. C. Pitkethly, Anal. Chem., 30:1309 (1958).
74.   J. C. Sternberg and R. E. Poulson, J. Chromat., 3:406 (1960).
75.   E. Evrard, M. Thevelin, and J. V. Joossens, Nature, 193:59 (1962).
76.   I. V. Markevich and S. D. Dobychin, Gas Chromatography, Proceedings of the
      Third All-Union Conference [in Russian], Izd. Dzerzh. Fil. OKBA (1966), p. 382.
77.   B. P. Okhotnikov, I. V. Bondarenko, and A. A. Datskevich, Gas Chromatography,
      Proceedings of the Third All-Union Conference [in Russian], Izd. Dzerzh. Fil.
      OKBA (1966), p. 377
78.   E. Cremer, T. Kraus, and E. Bechtold, Chem.-Ing.-Techn., 33:632 (1961).
79.   G. G. Devyatykh, N. Kh. Agliulov, and V. V. Luchinkin, Zav. Lab., 29:901
      (1967).
80.   E. Bechtold, Z. Anal. Chem., 221:262 (1966).

Chapter V

Nonisothermal Methods in Impurity Analysis

Various programmed-temperature methods are used in impurity analysis by gas chromatography.

Numerous methods have been described in which the temperature varies in time and (or) along the column, and there are commercial instruments giving reliabl: separations with set temperature programs.

There are detailed surveys [1, 2] and reviews [3-5] of chromatographic separation under nonisothermal conditions, so here we consider especially the scope for using the thermal factor in impurity analysis.

This approach allows one to increase the peak concentration in a band and hence to obtain a better limit of detection. Sometimes concentration and separation can conveniently be combined. Also, the analysis time is much reduced, and one can test for substances that differ greatly in boiling point.

Chromatography with temperature programming may be considered as a combination of concentration by thermal sorption with chromatographic separation.

Temperature programming is the most widely used method [6], in which the temperature of the whole column is raised during the separation. This allows the initial temperature to be lower without increasing the total duration of the analysis, while providing better separation of the volatile components. The completion at

high temperatures reduces the peak widths and limits of detection
for slow components.

The precise retention is dependent on the temperature program, the column characteristics, the temperature coefficient of
the partition factor, the gas flow rate, etc. See [1, 4, 7-10] on theoretical calculation of the retention.

It is more complicated to determine quantitatively the impurity composition under these conditions, so the method is best used
with mixtures whose qualitative impurity composition is known.
The method is especially good in purity analysis for compounds
whose impurities differ greatly in retention time. For instance,
it has been said [1] that the best separation of two closely spaced
peaks is probably obtained under isothermal conditions, but temperature programming can improve the degree of separation for
substances that differ widely.

The method is best used when the range in the boiling points
of the components exceeds 50-100°C, and sometimes when the
range is smaller, e.g., when selective sorbents are used or rapid
analysis is needed.

The peak concentration is increased in temperature programming and can be estimated roughly from

$$c_T \approx c_{T_0} \frac{\Gamma_{T_0}}{\Gamma_T} \approx c_{T_0} \exp\left(-\frac{\Delta H}{R} \cdot \frac{\Delta T}{T_0 T}\right), \qquad (48)$$

where c_T is the peak concentration with temperature programming,
c_{T_0} is the same at constant temperature T_0, and Γ_T and Γ_{T_0} are
the partition coefficients applicable to those two conditions, ΔH is
the sorption enthalpy, and T is the temperature at which the substance is eluted in programming.

Equation (48) shows that c_T is related to ΔH and ΔT. Zhukhovitskii and Turkel'taub [14, 15] were among the first to point
out that temperature programming can give an improved limit of
detection in gas chromatography.

Datskevich et al. [16] discussed enrichment of slow and fast
components by temperature programming, and they showed that
the peak output concentration under these conditions can substantially exceed the concentration in the initial sample (Table 7).

TABLE 7. Hydrocarbon Enrichment
by Temperature Programming [16]

Component	% in initial sample	Degree of enrichment	c_{max}, %
Butane	0.0004	0.013	32
Isopentane	0.0005	0.012	24
Hexane	0.0005	0.012	24

Janák [17-19] made a major contribution to the use of this method, while Juranek [20] devised methods for impurity analysis on microsamples.

Temperature programming reduces the analysis time and allows one to use a single column to analyze for impurities that boil over a wide temperature range [21-23].

The main advantage of temperature programming in impurity analysis is that the temperature rise elevates the gas-phase concentration, i.e., improves the sensitivity.

A major disadvantage of this approach is that the separation performance is less than that in the isothermal method. The thermal pulse method [24, 25] reduces this undesirable effect for a mixture with a relatively wide range of retention times. After an isothermal separation, the column is heated rapidly to a temperature at which the slowest component is hardly retained at all. This stops the separation, and all the components already separated are rapidly eluted.

This method narrows the impurity bands considerably, with a corresponding increase in the gas-phase concentration. The peak concentration for a fast component is increased by a factor $\Gamma_I (t_i /t_s)^{1/2}$, while that for a slow component is increased by a factor Γ_{II}; here Γ_I and Γ_{II} are the partition coefficients, t_i is the analysis time under isothermal conditions, and t_s is the same with shock heating.

The isothermal and programmed-temperature methods may be compared via the separation attained in a given time.

The pulse method allows one to use a longer column with the same analysis time as for the isothermal method. The fast components travel a long way down the column before the sharp rise, so their separation is increased. A simple theory gives the increase in the separation factor for the fast components from

$$K^{tp} = K^{iso} \sqrt{\frac{t_s}{t_i}}, \tag{49}$$

where K^{tp} is the separation factor for the thermal pulse method and K^{iso} is the same for the isothermal method.

The separation of the slow components is almost unaltered, but a major point in impurity analysis is the improved separation for the fast components, since (see Chapter I) the excess of the main component reduces the separation performance precisely for those components.

The pulse method has several advantages over temperature programming: (1) the enrichment factor is much larger (especially for fast components), since the elution is performed at a high temperature (when even the slowest component is scarcely held up), which is not possible with temperature programming, where the elution is performed at much lower temperatures (otherwise separation would not be attained); (2) temperature programming reduces the separation performance, and so the pulse method, which employs separation under the optimal isothermal conditions, provides better separation.

Figure 19 shows curves for impurities in isoprene as recorded under isothermal conditions and with rapid heating, the analysis time being the same in both cases. The second method gives larger peak heights (the limit of impurity detection is reduced by a substantial factor), while the separation is improved somewhat.

Chromathermography is [2, 26-31] another use of the thermal factor in chromatography, in which a temperature distribution is swept along the column, i.e., the temperature change is not simultaneous along the column. The carrier gas and heated zone move in the same sense in stationary chromathermography, and the temperature gradient in the oven is negative, i.e., the temperature falls in the direction of motion of the oven.

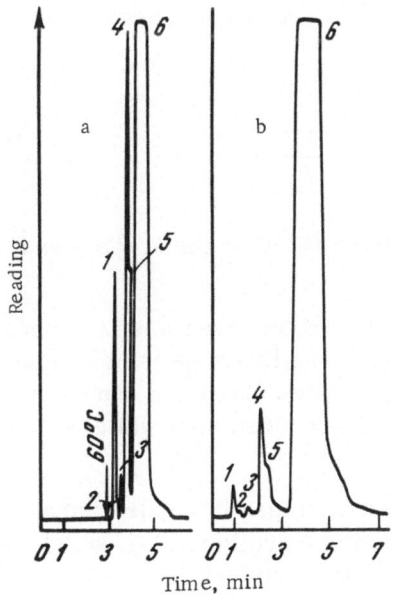

Reading

Fig. 19. Determination of impurities in isoprene: a) thermal-pulse method (isothermal separation at 0°C, rapid heating to 60°C), b) isothermal separation at 30°C; 1) 3-methylbut-1-ene, 2) pent-1-ene, 3) pent-2-ene, 4) 2-methylbut-1-ene, 5) 2-methylbut-2-ene, 6) isoprene.

Time, min

The mixture is separated into its components in the oven zone. The width of a band is restricted by the different speeds of the molecules at the edges of the band, on account of the temperature gradient. The mean speed for a linear isotherm corresponds to

$$V = \frac{\alpha}{\Gamma} = \frac{\alpha}{A \exp(Q/RT)}, \qquad (50)$$

where α is the linear velocity of the carrier gas, Q is the heat of sorption, T is the absolute temperature of the sorbent, and A is a constant for a given system.

From (50), molecules that run ahead of the band for any reason encounter cold sorbent, and their speed along the column is reduced. Conversely, molecules that lag behind encounter sorbent hotter than that at the band center, and there their speed is higher, so they soon catch up with the narrow band. A moving oven with a temperature gradient causes the compounds to group into bands having characteristic temperatures T_i provided that the Henry coefficient is substantially larger than the proportion of gas phase in the column:

$$T_i = - \frac{Q}{R} \cdot \frac{1}{\ln \chi \ \Gamma \frac{\omega}{\alpha}}, \tag{51}$$

where χ is the proportion of the column filled by the sorbent and ω is the speed of the oven.

This method produces symmetrical peaks even when the sorption isotherm is nonlinear, because the trailing edge is at a higher temperature than the leading one. Also, the maximum concentration is higher. Narrow columns of fine-grained material also help to produce narrow peaks [39].

Although this method has these advantages in impurity analysis, it is used less often than temperature programming, which we consider to be due mainly to the greater complexity of the equipment.

Another technique [32, 40] is separation simultaneously through out the column with a negative temperature gradient; a constant gradient along the column is used together with temperature programming. A fixed gradient gives no advantages over isothermal separation [2], but time variation in the temperature with a negative gradient is roughly equivalent to the temperature distribution in chromathermography, and this allows the latter method to be operated with a column of any length and shape.

Figure 20 shows a test on the method for impurities in toluene, and also the result from ordinary temperature programming. The new technique gives narrower peaks and improved separation of the impurities from the base. The maximum concentrations are 10-15 times those for isothermal separation.

Dantsig [33] described a pulsed thermal method for gas analysis. A high degree of enrichment was obtained in impurity analysis by separating the bands on one part of the column and compressing them on another. The process involves three stages: preliminary separation, band compression (with a narrow oven moving against the gas flow), and elution with temperature programming. It was found possible to assay ethane and ethylene in air at $10^{-4}\%$ in this way.

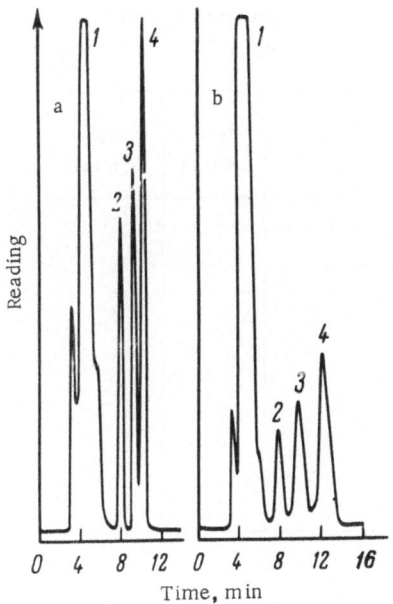

Fig. 20. Separation of impurities in toluene by: a) modified chromathermography, b) temperature programming: 1) toluene, 2) nonane, 3) mesitylene, 4) decane. Glass column 35×0.4 cm (gradient in case a produced by an external heater spiral of varying pitch), 10% Apiezon L on Spherochrome 1, nitrogen 15 ml/min, temperature rise 10 deg/min.

Skornyakov [34] has described an interesting thermal pulse method that avoids loss of separating capacity. The bands are compressed after the components have been separated under isothermal conditions, for which purpose they are passed into an additional short column kept at a lower temperature. The column is then rapidly heated, and the components are flushed into the detector by a gas flow. The compression ratio (enrichment factor) equals the ratio of the Henry coefficients for the component at the two temperatures of the short column.

An advantage of temperature programming is that the separation performance is only slightly dependent on the sample size, so an analysis for slow impurities can be performed by injecting a large sample (up to 500 μl [35]) with the column at a low temperature, when the V_R are high. The column then acts also as a concentrator. It has been pointed out [36] that this is a simple and effective method of concentration. However, it is sometimes more convenient to use a short forecolumn as the concentrator [37], especially in concentration at low temperatures. Impurities are best separated from the base (e.g., solvent) by means of a side-tube between the forecolumn and main column [38].

TABLE 8. Comparison of Isothermal and Other
Methods in Impurity Analysis

Characteristic	Iso-thermal	Noniso-thermal
Separation of components with similar properties	+	
Separation of components differing in properties		+
Limit of detection		+
Analysis of large samples		+
Wide choice of immobile phases	+	
Low cost	+	
Note: + indicates good (preferred use).		

Table 8 compares the advantages and disadvantages of isothermal and other methods as relevant to impurity analysis. Of
course, a particular problem may make it advantageous to use both
approaches. The table shows that nonisothermal methods are generally desirable in impurity analysis, while sometimes it is best
to combine the two approaches.

Future research on nonisothermal methods should provide
the basis for new effective methods and should serve to define the
areas of best use for existing methods.

LITERATURE CITED

1. W. E. Harris and H. W. Habgood, Gas Chromatography with Temperature Pro-
    gramming [Russian translation], Moscow, Mir (1968).
2. A. A. Zhukhovitskii and N. M. Turkel'taub, Gas Chromatography [in Russian],
    Moscow, Gostoptekhizdat (1962).
3. L. M. Kontorovich and V. B. Bobrova, Zav. Lab., 29:1027 (1968).
4. G. Ya. Myakishev, Usp. Khim., 36:1484 (1967).
5. J. Gas Chromat., 2:202 (1964); 3:210 (1965).
6. J. Griffiths, D. H. James, and C. S. G. Phillips, Analyst, 77:897 (1952).
7. H. W. Habgood and W. E. Harris, Anal. Chem., 32:450 (1960).

8. F. Baumann, R. T. Klaver, and J. F. Johnson, Gas Chromatography, edited by M. van Swaay, London, Butterworths (1962), p. 152.
9. M. J. E. Golay, L. S. Ettre, and S. D. Nogare, Gas Chromatography, edited by M. van Swaay, London, Butterworths (1962), p. 139.
10. S. D. Nogare and W. E. Langlois, Anal. Chem., 32:767 (1960).
11. J. C. Giddings, J. Chromat., 4:11 (1960).
12. R. Rowen, Anal. Chem., 33:510 (1961).
13. R. Rowen, Anal. Chem., 34:1042 (1962).
14. A. A. Zhukhovitskii and N. M. Turkel'taub, Usp. Khim., 25:859 (1956).
15. A. A. Zhukhovitskii and N. M. Turkel'taub, Usp. Khim., 26:992 (1957).
16. A. A. Datskevich, A. A. Zhukhovitskii, and N. M. Turkel'taub, Gas Chromatography, Proceedings of the Second All-Union Conference [in Russian], Moscow, Nauka (1964), p. 44.
17. J. Janák, Chem. Listy, 47:464 (1953).
18. J. Janák, Coll. Czechosl. Chem. Commun., 20:336 (1955).
19. J. Janák, Coll. Czechosl. Chem. Commun., 19:700 (1954).
20. J. Juranek, Theory and Applications of Chromatography [Russian translation], Moscow, Izd. AN SSSR (1960), p. 323.
21. L. M. Kontorovich and L. V. Andreev, Zav. Lab., 31:5 (1965).
22. V. V. Alekseeva, V. P. Bobrova, and A. I. Fomina, Neftekhimiya, 5:449 (1965).
23. G. A. Dantsig, Candidate's Thesis, Moscow, GIAP (1968).
24. J. Jacobs, Anal. Chem., 38:43 (1960).
25. V. G. Berezkin and V. S. Tatarinskii, Neftekhimiya, 6:492 (1966): Author's certificate 177,679 (1965); Byull. Izobr., No. 1, 105 (1966).
26. A. A. Zhukhovitskii, O. V. Zolotareva, V. A. Sokolov, and N. M. Turkel'taub, Dokl. AN SSSR, 77:435 (1951).
27. A. A. Zhukhovitskii, N. M. Turkel'taub, and V. A. Sokolov, Dokl. AN SSSR, 88:859 (1953).
28. N. M. Turkel'taub, V. P. Shvartsman, T. V. Georgievskaya, O. V. Zolotareva, and A. I. Karymova, Zh. Fiz. Khim., 27:1827 (1953).
29. A. A. Zhukhovitskii and N. M. Turkel'taub, Dokl. AN SSSR, 94:77 (1954).
30. A. A. Zhukhovitskii, N. M. Turkel'taub, and V. P. Shvartsman, Zh. Fiz. Khim., 28:1901 (1954).
31. K. W. Ohline and D. D. De-Ford, Anal. Chem., 35:227 (1963).
32. V. R. Alishoev, V. G. Berezkin, and V. S. Tatarinskii, Authors' certificate 218,513 (1967); Byull. Izobr., No. 17, 100 (1968).
33. G. A. Dantsig, Zav. Lab., 30:1313 (1964).
34. É. P. Skornyakov, Authors' certificate 171,656 (1963); Byull. Izobr., No. 11, 103 (1965).
35. L. Hollingshead, H. W. Habgood and W. E. Harris, Canad. J. Chem., 43: 1560 (1965).
36. J. J. Kirkland, Anal. Chem., 34:428 (1962).
37. J. Hornstein and P. E. Growe, Anal. Chem., 34:1354 (1962).
38. R. Abel, J. Chromat., 13:14 (1964).
39. V. Cantuti and G. P. Cartoni, Chim. e Ind., 50:449 (1968).
40. V. G. Berezkin, V. S. Tatarinskii, and A. A. Zhukhovitskii, Zav. Lab., in press.

Reactive Analytical Gas Chromatography

This is a recent development, which has in part been applied in impurity analysis. It is a form of gas chromatography in which the system includes a reaction unit [1, 2].

The first measurements involving reactions were made in 1955 [3-5], but the major advances have been made only in the last decade. The earliest studies in this area were concerned with impurities masked by the base [3] and with enhancing detector response [4].

The new method provides much greater scope than either of the two parent methods, since it gives information about the chromatographic and chemical properties of the compounds simultaneously.

The main objects are to simplify analysis and to extend the range of application. The column performance and detector response usually remain unchanged, but the reaction produces new compounds differing in separation and detector response, which thus imply changes in the basic chromatographic parameters.

The method has the following potential advantages over classical ones:

1. The range of application is increased, and unstable or nonvolatile compounds can be examined;
2. Peak identification is simplified, e.g., by the use of group reactions;

TABLE 9. Methods of Impurity Anal-

Compound	Changes in chromatographic characteristics	
	Retention time increased	Retention time decreased
Base	1. Separation of impurities from a base that gives a compound of low volatility [3, 6] 2. Frontal chemical concentration [7, 9]	
Impurity	3. Impurity concentration by use of chemical absorbers that form nonvolatile compounds with the impurities [10, 11]	4. Separation of impurities and base by converting impurities to volatile compounds by the use of: (a) tubular reactor [12] (b) liquid bubble-tower reactor [13, 14]
Carrier gas		

3. The actual performance is improved, because peak overlap can be eliminated;
4. There is an improvement in impurity detection.

Novel techniques and apparatus are involved. Here we consider only techniques for solving the specific problems of impurity analysis, since there is a good survey [2] of the basic techniques and regions of application.

The techniques sometimes provide simple solutions to complex problems, e.g., when the base completely overlaps the impurity peaks. Table 9 gives the general techniques that have

ysis in **Reactive Gas Chromatography**

Changes in detection characteristics	
Detection sensitivity increased	Detection sensitivity decreased
	7. Impurity detection against a background of a base that forms an undetected compound [19]
5. Conversion of undetectable impurities to compounds that are recorded by sensitive detectors by: (a) conversion in one stage [15] (b) conversion in two stages for compounds not containing carbon [16]	
6. Unselective concentration of impurity and base bands by chemical binding of part of carrier gas [17, 18]	

been developed for impurity analysis when the base differs in reactivity from the impurities.

Table 9 gives only the main techniques, although there are papers in which these techniques are used together [20].

Use of Reagents That Form Nonvolatile Compounds with the Base

Here the base forms a compound that is virtually nonvolatile under the working conditions (temperature, etc.) so that the impurities are completely separated from it. The reaction may be performed at any stage from injection to detection.

Ray [3] was the first to use this method to detect nonolefinic impurities in ethylene. The sample (10-25 ml) first entered a reaction tube (19 × 1.1 cm) filled with activated charcoal containing 40% bromine. The liquid bromination products from the ethylene were strongly retained by the charcoal at room temperature, while the nonolefins (permanent gases and saturated hydrocarbons) entered a stream of CO_2 and passed to the 40 × 0.2 cm chromatography column, which was filled with activated charcoal. The detector was a nitrometer containing alkali [21, 22]. Impurities at the 10^{-1} to $10^{-2}\%$ level were detectable. A more sensitive detector would undoubtedly improve on this.

Berezkin et al. [6] used a similar method to detect hydrocarbon contaminants on toluene. Nonpolar phases provide satisfactory separation of the impurities, but the toluene band overlapped some of the impurity ones. The toluene was removed selectively with a reaction system 20 cm long filled with diatomite soaked in concentrated sulfuric acid. The separation performance was hardly affected.

A chemically active sorbent (sulfuric acid on a carrier) has been prepared via a fluidized bed [23], which gives a uniform distribution of the acid over the diatomite, much reduces the preparation time, and allows the process to be conducted in a dry atmosphere, which is important because concentrated sulfuric acid is hygroscopic. Fluidized beds had previously been used to prepare column fillings in gas—liquid chromatography [24], and they appear especially useful in reactive gas chromatography when the reagent decomposes in air or it is difficult to choose a suitable solvent.

This technique provided for qualitative and quantitative analysis of toluene of AR, reagent, and scintillation grades; 18 impurities were identified. Figure 21 shows a chromatogram when sulfuric acid is used as the reagent.

If the reaction is slow, it is best to absorb the base under quasistatic conditions in a reactor in front of the column, e.g., in maleic anhydride in the analysis of buta-1,3-diene [25]. The sample (50-100 ml) enters the reactor at 100-1100°C in 30-60 sec.

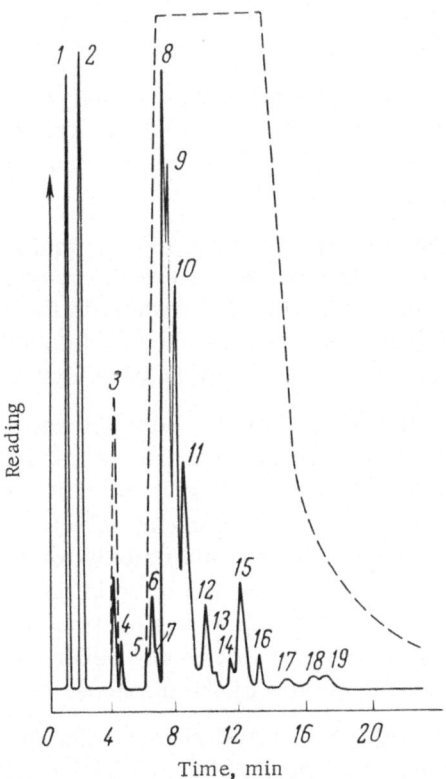

Fig. 21. Analysis of AR toluene on apiezon with use of a reactor containing concentrated H_2SO_4 (the broken line shows the peak given by benzene and toluene in ordinary analysis): 1) methane, 2) n-pentane, 3) benzene, 4) n-heptane, 5) methylcyclohexane, 6) ethylcyclopentane, 7) 3-methylheptane, 8) 1,2,4-trimethylcyclopentane, 9) 2-methylheptane + n-octane, 10) 4-methylheptane, 11) 1,1-dimethylcylcohexane, 12) trans-1,2-dimethylcyclohexane, 13) trans-1,3-dimethylcyclohexane, 14) cis-1,2-dimethylcyclohexane, 15) n-propylcyclohexane, 16) ethylcyclohexane, 17) n-nonane, 18) 1,1,3-trimethylcyclohexane, 19) cyclooctane.

The vessel contains kieselguhr bearing 30% maleic anhydride and 2.5% benzidine. The unreacted impurities are then separated on a sodium aluminosilicate column bearing 20% dimethylformamide. The following impurities were detectable: ethylene, propane, propylene, isobutane, n-butane, n-butylene + isobutylene, trans-butyl-2-ene, and cis-butyl-2-ene. Although impurities at the $2 \cdot 10^{-2}\%$

level were detectable, this limit could be reduced considerably, especially by the use of a more sensitive detector.

If reactive compounds are to be analyzed, the sorbent can often be a reagent that selectively takes up the base, as has been done in the analysis of BF_3 [26], which is strongly sorbed by a column containing dinonyl phthalate.

Hydrogen bonding to the stationary liquid can be considered as a particular case of reaction. Here the reagent can be quickly regenerated by heating [27] or by flushing in the reverse direction [28]. Hydrogen bonds can also be used to determine impurities [27] with a column that selectively retains higher alcohols.

The size of the reactor can be much reduced, and the most suitable liquid can be used to separate the impurities (in particular, one can use a liquid that does not give selective retention of the base), if the reactor contains a sorbent selective for the base but is connected in series with the column only for the time needed to isolate the impurities, the flow direction in the reactor being then quickly reversed while the impurities continue to be separated in the column. It is often desirable to use reversed flow in impurity analysis; see [29] for details of the methods.

Swoboda [28] used partial reverse flushing in the analysis of simple alcohols in aqueous solution. The gas flow after sample injection passed through a reaction vessel containing 20% diglycerol and a column containing 10% of polyethylene glycol 400; then, when the water zone was in the reactor but the relevant impurities had entered the column, the gas flows were switched to reverse the flow in the reactor while leaving that in the column unchanged.

This simple and effective method has the limitation that it cannot be used if the impurity has the same chemical nature and reactivity of the base.

There are difficulties in direct gas-chromatographic determination for impurities at the 10^{-5} to $10^{-8}\%$ level even when highly sensitive detectors are used. Analysis of special-purity compounds often requires preliminary impurity accumulation. Here the frontal method [30] (see Chapter VII for details) is effective. A form of this is frontal chemical concentration, where the sorbent is replaced by a reagent that reacts with the base to produce a strongly retained compound. This approach has the following advantages over use of a sorbent of the ordinary type:

1. It is possible to analyze for all unreacted impurities, because the product from the main component is not volatile and all impurities run ahead of it;
2. The reagent usually has a high capacity, so higher degrees of concentration are attained for a given vessel size;
3. The equilibrium constant may have a high temperature coefficient, so the base may be desorbed at relatively low temperatures, where most compounds are stable.

This method has been used [7] in analysis of hydrocarbon impurities in CO_2, with diethanolamine on reagent, which absorbs CO_2 reversibly but has very little effect on the impurities.

Figure 22 shows such an apparatus. The stainless-steel column 400 × 0.4 cm was filled with Inza diatomite bearing 30% diethanolamine. The chromatography column 120 × 0.2 cm was filled with ASK silica gel bearing 1.5% vaseline oil. The 25- to 250-ml sample of CO_2 passed to the reaction column via one part of the six-way stopcock, which removed the CO_2 [31], while the unreacted impurities formed a concentrated zone ahead of the CO_2 front, which was directed via the stopcocks to the chromatographic

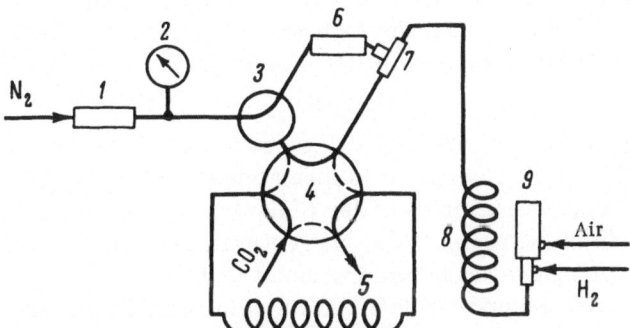

Fig. 22. Chromatographic system with reactor: 1) dryer for carrier gas, 2) precision gauge, 3) three-way stopcock, 4) six-way stopcock, 5) reactor, 6) resistance, 7) T-piece, 8) chromatography column, 9) flame-ionization detector.

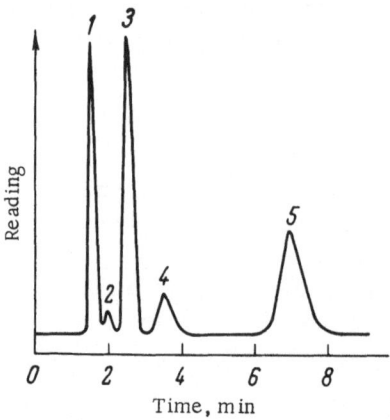

Fig. 23. Analysis of C_1-C_3 hydrocarbons in CO_2 (200 ml); 1) $7 \cdot 10^{-6}\%$ methane, 2) ethane, 3) $5 \cdot 10^{-6}\%$ ethylene, 4) $7 \cdot 10^{-7}\%$ propane, 5) $2 \cdot 10^{-6}\%$ propylene.

column. The flow rate of the carrier gas (60 ml/min) was almost unaffected by the stopcocks because the system contained a compensating resistance to balance out the effect of the reactor resistance. The system was operated at room temperature. The reactor was regenerated by heating to 100-105°C in a carrier-gas flow for 5-7 min. The heating and cooling were provided by equipment from a KhT-2M (Russian) chromatograph. A complete analysis cycle took 25 min. The method detects hydrocarbons in CO_2 at the $10^{-6}\%$ level (Fig. 23).

Direct elution analysis even of a large sample (25 ml) of this CO_2 failed to reveal the hydrocarbons. Similar methods can be used with contaminants in other acid gases (e.g., H_2S, HCN, etc.).

The scope of the method is governed by the available reactions. Chemical absorption [32] has been used [8, 9] in frontal chemical concentration in order to determine traces of carbon compounds in hydrogen. The reagent was palladium black, which has very pronounced chemisorption of hydrogen at room temperature (it absorbs up to 200 times its volume of hydrogen), and all impurities from CO to butane run ahead of it. Adsorption concentration of impurities from hydrogen causes great difficulty in the choice of a sorbent producing efficient adsorption at room temperature and also desorption at elevated temperatures for substances differing greatly in boiling point, e.g., CO and butane.

The hydrogen passes through a fine-control valve to the concentrator, which contains palladium black on asbestos (80×1 cm, capacity 1.5 liters of hydrogen). The concentrator had been flushed with an inert gas at an elevated temperature to remove hydrogen. Ahead of the hydrogen there arose a concentrated impurity band, which eventually passed to the 250×0.4 cm analysis column, which contained one of several possible materials: a molecular sieve to separate permanent gases, alumina with 3% NaOH to separate hydrocarbons, etc. In the latter case, a flame-ionization detector was used, while a katharometer was used in the former case. The carrier was the original hydrogen, whose flow rate was reduced to 50 ml/min before the impurity band broke through from the concentrator.

Figure 24 illustrates the effects of enrichment. The process also produces quantitative hydrogenation of unsaturated compounds to the corresponding hydrocarbons, while CO is converted to methane. The components were at the $4 \cdot 10^{-5}\%$ level in the initial mixture, while the lowest CO concentration that could be determined in this way was $10^{-8}\%$.

The absorption column was regenerated by flushing it first with air (with water cooling) and then with inert gas while it was heated to 100°C. One analysis cycle took 30 min.

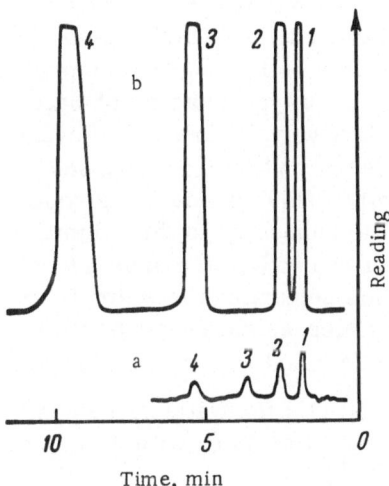

Fig. 24. Analysis of hydrocarbons ($4 \cdot 10^{-5}\%$) in hydrogen; a) without enrichment, b) enrichment on palladium black; a) 1) methane, 2) ethane, 3) ethylene, 4) propane; b) 1) methane, 2) ethane, 3) propane, 4) butane.

It is also possible to use a material without regeneration, in which case a small reactor is inserted in the flow as in Fig. 22. Impurities in permanent gases have [33] been determined with moist NaOH in the reactor.

Frontal chemical concentration can be used also for intermediate concentration of slow impurities (around $10^{-4}\%$) in ethylene and propylene via a short column containing diethanolamine [34], the carrier gas being pure CO_2. This combines three steps: preliminary separation, concentration of slow components, and analysis of the concentrate.

Selective retention of the main component is now very widely used in impurity analysis.

Impurity Concentration by Chemical Retention of the Impurities

Preliminary concentration is often a necessary stage in impurity analysis, especially for inorganic impurities, since sensitive and reliable detectors are not so readily available for these.

Adsorption is commonly used [35], but it is often restricted to strongly adsorbed (slow) components; also, there is the possibility of thermal decomposition during desorption (usually performed at high temperatures), and recovery may not be complete.

Chemical concentration is promising [10, 11] and uses selective formation of compounds of low volatility, which may be released unchanged or as reaction products on raising the temperature or applying fresh reagents. The prototype is assay of water in butane [36], where the water is selectively absorbed (via hydrogen bonds) in polyethylene glycol. The gas at 10°C was passed through a 30.5×0.63 cm trap containing 30% of grade 200 glycol. Then the trap was heated to 90°C while connected to the column with helium flowing; the column contained 30% of the same glycol on diatomite and was 61 cm long. This separated the water from the other possible impurities (methyl mercaptan, benzene, etc.). A concentration of $2 \cdot 10^{-5}\%$ could be detected on a 10-liter sample.

A similar method has been used [37] for water in isopentane, the trap containing triethylene glycol at 25°C, with desorption at 100°C and a limit of detection of $10^{-3}\%$.

These methods can be applied to water in many hydrocarbon gases (or liquids, if solvent extraction is used); but it appears to be more selective to determine water by the use of salts to form hydrates.

Traces of CO_2 and H_2S can be concentrated because acid gases form unstable compounds with organic bases (e.g., ethanolamine) at room temperature. These compounds readily decompose on raising the temperature. The isolated material is separated with a column.

Triethanolamine has been used in this way [38] with CO_2 and H_2S. The enrichment factor is [39]

$$K_o = \frac{c_f}{c_i} = \frac{K_i}{K_f} , \qquad (52)$$

where c_i is the CO_2 concentration in the initial mixture, c_f is the same in the final mixture, K_i is the partition coefficient at the initial temperature, and K_f is the same at the final temperature. This gave about 750 for K_0 for CO_2 with temperatures of 22 and 99°C. Diethanolamine gives an even larger factor, but this compound is not so suitable as it is fairly volatile and tends to be unstable under the working conditions.

The apparatus was similar to that of Fig. 22. The concentrator column was 36×0.4 cm and was filled with 30% triethanolamine on diatomite. The CO_2 and H_2S were separated with a column 450×0.4 cm [40] containing 25% hexadecane on diatomite. The concentrator was coupled to the analysis column via a six-way stopcock. The sample was 100-2000 ml. After the impurities had accumulated, the concentrator was flushed with carrier gas at 30-40 ml/min for 4-6 min to remove the major component; then the concentrator was rapidly placed in a bath containing boiling water and the released acid gases were directed to the analysis column, which worked into a katharometer. The carrier gas (30-40 ml/min) was hydrogen or helium.

This method gave a limit of detection of 10^{-4}% for acid gases in propylene. One analysis cycle takes 15-20 min. It is undesirable for very acid gases (SO_2 and SO_3, for instance) to be present, because they form stable salts with the triethanolamine and reduce its capacity. A method based on xylidine [10] has been used for SO_2.

Chemical concentration has the following advantages over ordinary sorption:

1. High selectivity (one impurity or a few may be accumulated),
2. The capacity is usually high (and hence also the degree of concentration),
3. The equilibrium constant has a higher temperature coefficient, so only a relatively small temperature rise is needed to release the impurities.

Chemical concentration followed by thermal desorption requires only simple apparatus, but it is restricted to reversible reactions. This makes it desirable to extend the method to a wider class of reactions. Special reagents are needed for this purpose.

A liquid wash-bottle provides a convenient way of performing the initial absorption and subsequent release reaction.

It has been shown [11] that this method has advantages in the case of preliminary accumulation of CO_2 and H_2S by NaOH solution; acidification releases these gases in concentrated form for passage to the analysis column. The system is similar to that of Fig. 22, except that the gas-washing system of Fig. 25 is used instead of an accumulation column.

The aqueous KOH is injected at the bottom. The device is fed via the appropriate arm of the six-way stopcock, and the gas volume is monitored by a GSB-400 meter. The sample is followed by an appropriate amount (twofold excess) of 10% sulfuric acid, and the released acid gases pass to a 150×0.4 cm column containing ASK silica gel bearing 1.5% of vaseline oil. Air, CO_2, and H_2S are quite well separated. The limit of detection is about $5 \cdot 10^{-4}\%$. One analysis takes 20-30 min.

A form of this method has been used [11] for traces of H_2S in gases. The sample is drawn through a glass tube (5×0.3 cm) filled with porous porcelain treated with lead carbonate. When most of the reagent has been used up, the porcelain is tipped onto a glass filter, through which hydrochloric acid (1:1) rises from below, and the released H_2S is examined by chromatography. The minimum detectable H_2S concentration was about $5 \cdot 10^{-4}\%$.

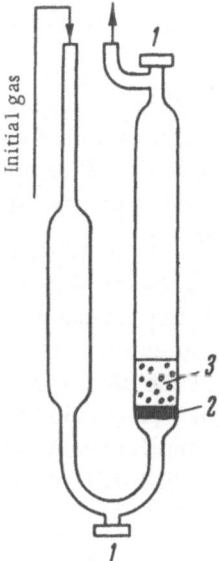

Fig. 25. Glass accumulation device:
1) upper and lower rubber injection
discs, 2) porous plate, 3) solution.

Volatile aldehydes and ketones may be concentrated via the semicarbazones followed by release with phosphoric acid [41]. Similar methods can be devised for other classes of compound.

Conversion of Impurities to Compounds Convenient for Assay

If an impurity produces a volatile compound readily separated from the base, a much improved limit of detection may be available, since the concentration of the eluted compound is increased in proportion to the reciprocal of the retention volume.

It is difficult to assay water directly in various technical solvents having similar V_R, and so Bayer [12] passed the sample in a carrier gas through a column containing calcium carbide, which reacts with water to give acetylene, which at room temperature elutes as a narrow peak well ahead of the complex solvent mixture. The method is applicable down to $10^{-3}\%$.

Knight and Weiss [42] made a more detailed study of the carbide method for water. The acetylene was recorded by a flame-ionization detector, which improved the limit of detection by almost an order of magnitude. A similar method has been used [43] for water in burnt hydrocarbon gases.

The carbide method involves certain kinetic difficulties, and so the reaction for water in anhydrous ammonia was performed [44] for 10 min in a vessel. It was not possible to get complete reaction in a flow in a 30×0.3 cm column filled with calcium carbide because the reaction is not rapid enough.

Drawert et al. [20] have used reactions to assay water in alcohol and ethanol in blood. In the first case, the reaction vessel was filled with a 1:1 mixture of Sterhamol and calcium hydride. The hydrogen was readily separated from all the alcohols as a narrow peak. In the second case, the ethanol was converted to ethylene in order to improve the separation and increase the response. Dehydration is quantitative at 200-300°C with phosphoric acid on Sterhamol (1:2).

Hydrogen may be assayed with a KhT-2M chromatograph after production from calcium hydride in the analysis of gases and liquids for water [45]. Traces of water can be detected [13, 14] in liquid hydrocarbons and some compounds containing oxygen by chromatography of the hydrogen released by sodium aluminum hydride in diethylene glycol dimethyl ether. Here a liquid bubbling unit was inserted in the system, which can be adapted to many reactions, including liquid-phase ones involving several reagents at room temperature, and it also greatly increases the sample volume that can be used, i.e., improves the limit of detection. Analysis for water in organic compounds is an important problem [36], and this method has advantages over published ones [42, 46-49] in being more sensitive and reliable for low water concentrations.

A unit of this type (Fig. 26) may be inserted in the system before the column. Carefully dried argon containing about 0.003% oxygen enters the unit from below and produces good mixing. The specimen is supplied from one burette, while the other burette contains a standard solution having a known water content.

There are two methods of determining the water content of the specimen: by comparison and from a calibration curve. In the latter case, the curve is drawn up from hydrocarbon specimens with known water contents, e.g., from solubility tables. The results

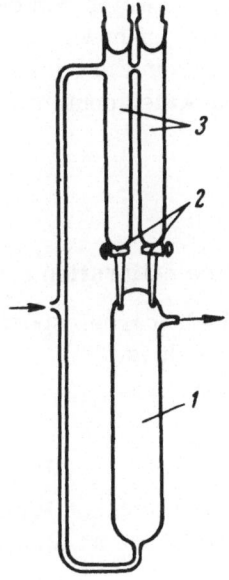

Fig. 26. Device for detecting traces of water in liquid samples: 1) reaction vessel, 2) stopcocks, 3) burettes for solutions

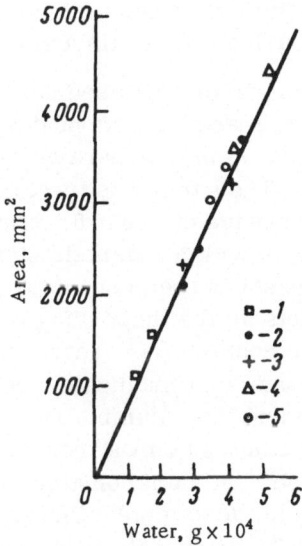

Fig. 27. Hydrogen peak area as a function of dissolved water content for: 1) cyclohexane, 2) benzene, 3) toluene, 4) xylene, 5) ethylbenzene.

are presented as peak areas as a function of the water content in a set volume (Fig. 27), which is used to deduce unknown water contents. If the volume W of the sample differs from the volume V of the standard used in the calibration, then the water content is given by

$$P = P_s \frac{V}{W},$$ (53)

in which P_s is the water content read from the calibration curve.

The comparative method employs two hydrocarbon specimens, one having a known water content, and the result is

$$P_x = P_s \frac{S_s V}{S_x W},$$ (54)

where S_s is the peak area given by the standard (content P_s) and S_x is the area given by the unknown, while V and W are as above.

The water contents of the solvents and monomers are important in polymer production. This method has been used with isoprene, styrene, heptene, vinylcyclohexane, etc., which are usually inert to lithium aluminum hydride [50]. The method has also been applied to water in ethers and cyclic esters [14].

The method has also been applied to water in inorganic materials such as minerals and metals.

Carpenter [51] used conversion to a volatile compound in rapid assay of traces of carbonates in inorganic materials, and he described a simple apparatus for attachment to a standard chromatograph. The sample is treated with 10 ml of $3N$ HCl with the system at a pressure of about 0.5 atm. The sample is stirred vigorously for 5 min with a magnetic stirrer during the reaction. The reaction vessel is then connected to a vessel of known volume, which also acts as the dispensing vessel for the chromatograph. The sample contains air, water, and CO_2. The water is removed in a tube containing magnesium perchlorate placed between the latter vessel and the column. The air and CO_2 are separated on a silica gel column (30 cm) at room temperature with a helium speed of 4 cm/sec. A katharometer was capable of determining 0.01 mmole of CO_2 in the sample. One analysis takes 10 min and can detect $0.2 \cdot 10^{-4}\%$ carbonate in the sample.

Jeffery and Kipping [52] independently described a method for assaying carbonate in rock. Dilute orthophosphoric acid was used. The same acid was used [53] in assaying CO_2 and NO in aqueous solutions of ethanolamine, with assay of the gases.

Juranek and Ambrova [54] had previously described a similar method for impurities in nonvolatile compounds. Carbon was assayed in the presence of sulfur for iron alloys and analogous materials by combustion in a stream of oxygen, which also acted as the carrier gas in the subsequent separation of CO, CO_2, and SO_2, the column output being monitored by a photocolorimetric cell of high sensitivity [55]. It was possible to detect $10^{-6}\%$ of carbon in a 1-g sample in an analysis lasting 25 min.

Walker and Kuo [56] described a sensitive and accurate method for carbon in iron alloys. The specimen was burned in a stream of oxygen in an induction furnace, and the gaseous products passed via a reactor containing manganese dioxide to a 1.2-m column containing 5A molecular sieve. After the sample had been burned, the CO_2 was eluted in helium with programmed temperature increase to 275°C. The peak area was determined with a disc diagram integrator. The method is applicable to samples with from $5 \cdot 10^{-4}$ to 20% carbon. One analysis takes 20 min.

Stuckey and Walker [57] described a method for simultaneous determination of carbon and sulfur in iron, which covered roughly the ranges from 0.01% to 10% carbon and from 0.01 to 0.1% sulfur. A description has been given [58] of a method of detecting $10^{-4}\%$ carbon in sodium. Another method has been described for traces of C and S in steel and iron [59], which employs thermal concentration of the CO_2 and SO_2; a 1-g sample allowed about $10^{-5}\%$ to be detected. Sukhorukov and Ivanova [60] used a flame-ionization detector to determine carbon in metals; the oxides of carbon were converted to methane.

Traces of carbon compounds in hydrogen peroxide can also be assayed [61], which is necessary in quality and stability control, since explosive mixtures can arise from the accumulation of carbon compounds. A 10-μl sample in a glass micropipette was inserted into the upper part of a quartz tube heated to 300°C. The sample then evaporated instantly, and a stream of helium carried the vapor to the lower part of the tube, which was heated to 900°C and was filled with quartz rods. The H_2O_2 decomposes and the organic compounds are oxidized. The products (water, oxygen, and CO_2) were carried by the helium through a tube containing magnesium perchlorate to the column, which was 60 cm long and 6 mm in diameter operated at 30°C.

The results were evaluated via a curve relating CO_2 peak height to carbon concentration. The range $1-200 \cdot 10^{-4}\%$ carbon was covered, but it has also been used up to 0.5% carbon. The coefficient of variation was about 5% for carbon contents of $100-200 \cdot 10^{-4}\%$. One analysis takes 15 min. The method replaces Lebich's much more complicated method.

Okuno et al. [62] used catalytic hydrogenation by platinum to assay S in organic compounds; the products (mainly methane and H_2S) were treated in cooled traps before analysis. Crude oil containing about $10^{-2}\%$ S was analyzed in this way, and it would seem [62] that the method might give access to $10^{-4}\%$.

It is sometimes best to convert a very reactive impurity to a less active one, e.g., Bond et al. [63] assayed traces of SO_3 in SO_2 by converting the former to oxides of carbon in a reactor filled with crystalline oxalic acid.

Diedrich et al. [64] determined $4 \cdot 10^{-4}\%$ ammonia after catalytic conversion to nitrogen.

Zorin et al. [65] converted hydrides to hydrogen at 1000°C in order to obtain stable katharometer readings.

Such conversions allow one to analyze selectively for volatile and other impurities in many compounds, as well as for very reactive impurities.

Adjustment of Detection Sensitivity via Reactions

The final result of a chromatographic analysis is determined not only by the separation in the column but also by the detector's selectivity. It is often possible to simplify an analytical problem considerably if one can use a reaction to give products from interfering compounds that do not affect the detector; alternatively, undetectable impurities may be converted to detectable ones.

We have seen in Chapter III that flame ionization is most widely used as a detector very sensitive to organic compounds and almost insensitive to inorganic ones such as water, oxygen, oxides of carbon, H_2S, etc. These latter can be detected at high sensitivity by converting them to organic compounds (e.g., methane) that can be detected in that way, especially since oxygen, water, etc., adversely affect polymerization, catalysis, etc., and so need to be assayed.

Schwenk et al. [15] used this method to detect traces of CO and CO_2 in ethylene. The components were separated on an activated-charcoal column in a stream of hydrogen and then passed to a 35 × 0.6 cm reactor filled with a nickel catalyst (10% nickel on a carrier) at 300°C, which converted the CO and CO_2 quantitatively to methane, which was then recorded by flame ionization. The limit of detection was 10^{-5}% for a 25-ml sample.

Porter and Volman [66] examined this conversion in detail for use in gas chromatography. The nickel catalyst was used in a 12.5 × 0.16 cm stainless-steel vessel. They found that the degree of conversion did not alter between 206 and 266°C, but the asymmetry and width of the methane peak were altered fairly substantially. It is best to convert CO to methane at 266°C, since this gave the most symmetrical peak. The hydrogenation was quantitative under these conditions, and the areas of the CO and CH_4 peaks given by 10-μl samples were virtually identical. It was also suggested that traces of hydrogen could be detected by flame ionization via conversion to methane with excess CO.

This reaction is widely used in assaying traces of carbon oxides in monomers [15, 67], in air [68], in mine gas [69], and in other industries.

The method has been extended by using reaction in several stages, and in this way it has been extended to compounds not containing carbon. For instance, oxygen can undergo double conversion via CO to the equivalent quantity of methane, which is then detected by flame ionization.

Quantitative conversion of combined oxygen to CO at carbon is widely used [70] in microanalysis for oxygen in organic compounds. The precise reaction conditions are important: type of carbon, temperature, etc. It has been suggested [71] that the best carbon for quantitative conversion is anthracene black from the Kadievo plant, while platinized black at 900°C reduces side-reactions between the black and the walls of the quartz tube [72]. Allowance was made for the conditions for quantitative conversion of CO to methane, which had been determined previously [15, 66, 67].

The oxygen is separated in the usual way on a column and then passes to the first converter, where it is quantitatively converted to CO at platinized carbon (50% Pt on Kadievo anthracene black). The CO then passes to the second converter, where it is transformed to methane in hydrogen over nickel; the hydrogen is introduced before the converter. The carbon (120-mm layer) in the first converter is bounded by quartz on both sides and is heated for 8 hr in a stream of pure argon before use. The second converter has a 120-mm layer of nickel catalyst. The column (200×0.4 cm) contained ASK silica gel. The area of the methane peak increased linearly with the amount of oxygen. The coefficient of variation was 5%, and it was possible to measure $5 \cdot 10^{-4}\%$ oxygen.

A sensitive method has been described [73] for COS and CS_2 in gases. These compounds are first isolated with a column containing silica gel (the other components are methane, carbon oxides, ethylene, acetylene, etc.) and then are hydrogenated at 280-540°C to methane over a nickel catalyst. The method has been applied to these compounds in monomers, and success was also reported [73] in preliminary tests on analysis for HCN and $(CN)_2$ by hydrogenation to methane and flame-ionization detection.

A new sensitive detector [74] employs reaction of the eluted compounds with a nonvolatile radioactive compound to give a volatile one that is recorded by a Geiger counter. The device has been used to detect traces of bromine, fluorine, chlorine, nitrosyl chloride, etc., in gases by reaction with the hydroquinone clathrate compound of ^{85}Kr (radioactive), which is essentially stable at room temperature. A nearly linear relation was obtained between the fluorine content and the count rate. A similar method can be used for traces of SO_2 [75]. See [76, 77] on the calibration of instruments with a radiation detector of this type.

There are several papers [78-80] on detectors employing Belstein's reaction in order to identify organic halogen compounds in chromatograph outputs. The flow (or part of it) passes after the katharometer through a burner whose flame impinges on a copper grid, which produces a green color in response to halogens. This color is also produced by organic compounds containing −CN, −CCN, etc. The limits of detection are $5\text{-}80 \cdot 10^{-4}\%$ [80].

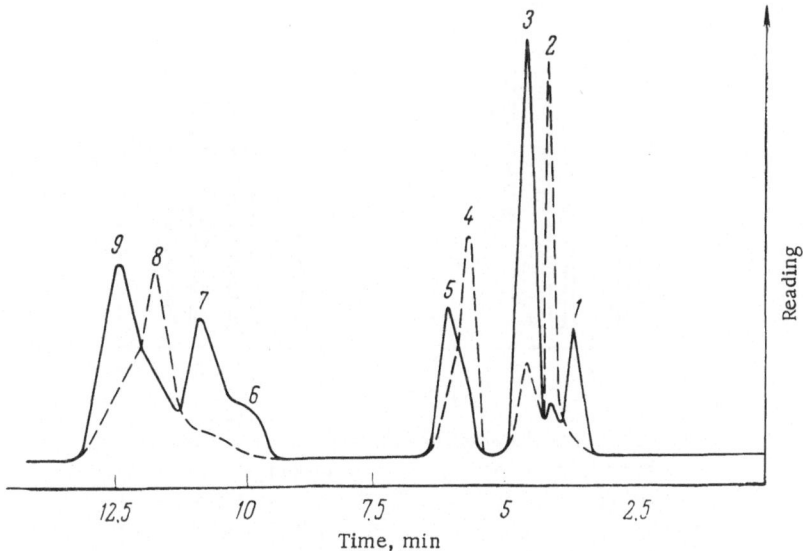

Fig. 28. Broken line recorded with a detector employing Belstein's reaction; solid line recorded by flame ionization: 1) 1,4-dimethylpentane, 2) 1,2-dichloroethane, 3) benzene, 4) 1,2-dichloropropane, 5) n-heptane, 6) 3,4-dimethylhexane, 7) 1,3-dimethyl-cis-cyclohexane, 8) 1,2-dichlorobutene, 9) n-octane. Stationary phase 30% SF-96/1000, 2-m column, 5.3 mm in diameter, 4 liters/hour of gas.

Figure 28 shows curves recorded in this way and by flame ionization. The selective detector records only the halogen compounds, and its response approximates to that of the flame-ionization detector, which records all organic compounds. The selective detector provides qualitative identification of trace impurities and also content assay of compounds of this class even when they are not chromatographically separated from other organic compounds.

Sometimes in impurity analysis it is desirable to adjust the detection sensitivity for the main component rather than the trace ones. Selective conversion of the base to an undetected compound provides access to impurities whose presence is usually concealed [19], e.g., for traces of C_1-C_4 hydrocarbons in oxides of nitrogen, which is important in the manufacture of nitrogen fertilizers, in the production of certain rarer metals, and in nuclear engineering.

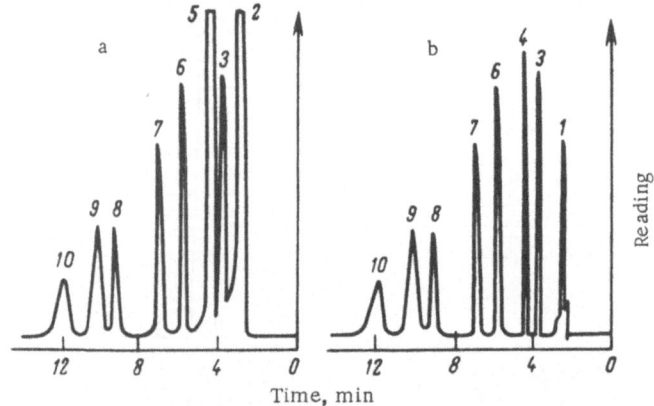

Fig. 29. Traces of hydrocarbons in nitrogen oxides assayed:
a) without hydrogenation, b) with hydrogenation: 1) methane,
2) methane + NO, 3) ethane, 4) ethylene, 5) ethylene + N_2O,
6) propane, 7) propylene, 8) isobutane, 9) n-butane, 10) n-
butylene.

The program heating unit from a KhT-2M was used with a
350×0.4 cm column filled with alumina containing 3% NaOH. Figure 29 shows a recording of traces of hydrocarbons in NO + N_2O.
In the 6-min analysis, the column temperature rose to 120°C, which
hardly affected the readings of the flame-ionization detector. The
determination is complicated by the fact that the detector responds
to NO and N_2O, whose peaks obscure some of the impurity ones.

It is well known that this detector responds to NO and N_2O [81],
so these were converted in a stream of hydrogen to water and nitrogen, which do not affect this detector, by passage over a palladium catalyst in a tube 20×0.4 cm made of stainless steel, which
was placed between the column and the detector, which had a platinum electrode. Figure 29 shows the result of doing this; the nitrogen oxide peaks are almost absent. The method gives access to
10^{-4}% hydrocarbon in the nitrogen oxides. An analysis takes about
10 min.

The response to trace components can be increased by removing some of the carrier gas after the column [17, 18], e.g., if a mix-

ture of helium and CO_2 is used [18], the latter being absorbed by alkali. This selective absorption reduces the band widths and correspondingly raises the peak concentrations. The entire gas flow is passed through a nitrometer (height 25 cm, volume 100 ml) filled with 40% caustic potash. The gas is then dried by a tube containing anhydrone mixed 1:1 with an inert carrier before passing to the conductivity detector. The initial flow rate was 100 ml/min. Figure 30 illustrates the separation of but-1-ene, cis-but-2-ene, and trans-but-2-ene for 0.04 ml samples when the carrier is (a) pure helium and (b) He + CO_2 (1:10). The separation was performed with a 400×0.4 cm column containing 25% hexadecane on Sterhamol. The peak heights in Fig. 30b have been increased by a factor 11. The alkaline solution did not alter the amounts of material in the peaks, so we have

$$c_i \mu_i = c_f \mu_f, \tag{55}$$

where c_i is the peak concentration before the absorber, c_f is the same after it, and the μ are the corresponding widths (in volume terms). As

$$\mu_f = (1 - \alpha) \mu_i \tag{56}$$

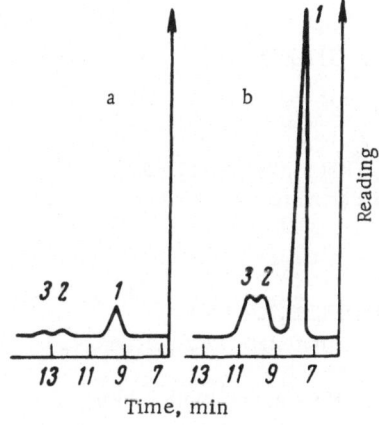

Fig. 30. Analysis of mixed butenes: a) without accumulation, b) with accumulation: 1) but-1-ene, 2) cis-but-2-ene, 3) trans-but-2-ene.

Time, min

(where α is the proportion of the gas absorbed), we get the follow-ing result:

$$K_0 = \frac{c_f}{c_i} = \frac{1}{1-\alpha}. \tag{57}$$

This agrees well with experiment. For instance, $\alpha = 0.91$ for a 1:10 He–CO_2 mixture, and (57) gives $K_0 = 11.1$, whereas experiment gave $K_0 = 11.0$. The time widths of the peaks are almost un-altered in this method. Tests have shown that the absorber in-creased the butene widths by not more than 5%.

This is a fairly general method of increasing concentrations. For instance, hydrogen as a component could be removed by pal-ladium black on asbestos or palladium diffusion tubes. One could also use a mixture whose components react with volume reduction. This unselective method could be used in chromato-mass spectrom-etry as a molecular separator.

* * *

The above examples illustrate the prospects for using reactive gas chromatography in impurity analysis.

At present, progress in methods in reactive gas chromatogra-phy has run ahead of apparatus development; but soon one expects to find chromatographs with a set of standard reaction vessels as common as chromatographs with a set of standard detectors now are.

LITERATURE CITED

1.   V. G. Berezkin and O. L. Gorshunov, Usp. Khim., 34:1108 (1965).
2.   V. G. Berezkin, Analytical Reactive Gas Chromatography [in Russian], Moscow, Nauka (1966).
3.   N. H. Ray, Analyst, 80:853 (1955).
4.   G. E. Green, Nature (London), 180:295 (1957).
5.   A. E. Martin and J. Smart, Nature (London), 175:422 (1950).
6.   V. G. Berezkin, B. V. Strizhkov, and V. S. Tatarinskii, Neftepererabotka i Neftekhimiya, No. 8, 26 (1967).
7.   V. G. Berezkin and O. L. Gorshunov, Izv. AN SSSR, Ser. Khim., 2069 (1965).

8. V. S. Mirzayanov, V. G. Berezkin, and A. A. Datskevich, Authors' certificate 171,660 (1964); Byull. Izobr., No. 11, 105 (1965).

9. V. S. Mirzayanov and V. G. Berezkin, Analysis and Monitoring Methods in the Chemical Industry [in Russian], No. 1, Moscow, NIITÉKhim (1966), p. 28.

10. V. G. Berezkin and O. L. Gorshunov, Zh. Anal. Khim., 21:1487 (1966).

11. V. G. Berezkin, O. L. Gorshunov, and Z. P. Markovich, Zav. Lab., 29:1067 (1967).

12. E. Bayer, Angew. Chem., 69:732 (1957).

13. V. G. Berezkin, A. E. Mysak, and L. S. Polak, Khimiya i Tekhnologiya Topliv i Masel, No. 2, 67 (1964).

14. V. G. Berezkin, A. E. Mysak, and L. S. Polak, Neftekhimiya, 4:156 (1964).

15. U. Schwenk, H. Hachenberg, and M. Förderrenther, Brenstoff-Chem., 42:295 (1961).

16. V. G. Berezkin, A. E. Mysak, and L. S. Polak, Izv. AN SSSR, Ser. Khim., 1871 (1964).

17. K. A. Gol'bert, O. L. Gorshunov, and V. G. Berezkin, Authors' certificate 193,777 (1966); Byull. Izobr., No. 7, 112 (1967).

18. K. A. Gol'bert, O. L. Gorshunov, and V. G. Berezkin, Zav. Lab., 29:799 (1967).

19. V. S. Mirzayanov, V. G. Berezkin, and V. A. Nikol'skii, Zh. Anal. Khim., 21:1239 (1966).

20. F. Drawert, K. Felgenhaner, and G. Kupfer, Angew. Chem., 72:385 (1960).

21. J. Janák, Chem. Listy, 47:464, 700, 817 (1953).

22. D. A. Vyakhirev, A. I. Bruk, and S. A. Guzlina, Dokl. AN SSSR, 90:577 (1953).

23. V. G. Berezkin and V. P. Pakhomov, Authors' certificate 166,850 (1964); Byull. Izobr., No. 23, 65 (1964).

24. V. G. Berezkin and V. P. Pakhomov, Khimiya i Tekhnologiya Topliv i Masel, No. 9, 58 (1965).

25. J. Janák and J. Novak, Coll. Czechosl. Chem. Commun., 24:384 (1959).

26. N. M. Turkel'taub, S. A. Ainshtein, and S. V. Vyavtsillo, Zav. Lab., 28:141 (1962).

27. R. G. Ackman and R. D. Burgher, J. Chromat., 6:541 (1961).

28. P. A. T. Swoboda, Chem. and Ind., 1262 (1960).

29. K. A. Gol'bert and M. S. Vigdergauz, Textbook of Gas Chromatography [in Russian], Moscow, Khimiya (1967).

30. V. S. Mirzayanov, A. A. Zhukhovitskii, V. G. Berezkin, and N. M. Turkel'taub, Zav. Lab., 29:1166 (1963).

31. R. Bottomsa and W. Wood, Refiner and Natural Gasoline Manufacturer, 3:105 (1935).

32. E. Glückauf and G. P. Kitt, Vapour Phase Chromatography Symposium, London, Butterworths (1957), p. 422.

33. D. G. Timms, H. J. Konrath, and R. C. Chirnside, Analyst, 83:600 (1958).

34. V. G. Berezkin and O. L. Gorshunov, Gas Chromatography, No. 5, Moscow, NIITÉKhim (1967), p. 77.

35. S. D. Nogare and R. S. Juvet, Gas—Liquid Chromatography, Theory and Practice (1962).

36. A. A. Cerbstrom, C. F. Spencer, and J. F. Johnson, Anal. Chem., 32:1056 (1960)

37. V. G. Baranov, A. G. Pankov, and Ya. I. Tur'yan, Principles of Physicochemical Methods of Analysis and Monitoring in Isoprene Production [in Russian], Moscow, NIITÉKhim (1965).

38. A. L. Cowl and F. S. Riesenfeld [in Russian], Gas Purification, Moscow, Gostekhizdat (1962).

39. A. A. Zhukhovitskii and N. M. Turkel'taub, Gas Chromatography [in Russian], Moscow, Gostoptekhizdat (1962).

40. S. A. Ainshtein, B. I. Anvaer, and N. M. Turkel'taub, Zav. Lab., 30:669 (1964).

41. I. R. Hunter and M. R. Walden, J. Gas Chromat., 4:246 (1966).

42. H. S. Knight and F. T. Weiss, Anal. Chem., 34:749 (1962).

43. A. Goldup and M. T. Westaway, Anal. Chem., 38:1657 (1966).

44. G. O. Guerrant, Anal. Chem., 39:143 (1967).

45. I. M. Starshov, Yu. V. Voevodkin, and A. A. Khalmov, Gas Chromatography, No. 3, Moscow, NIITÉKhim (1965), p. 103.

46. J. Mitchell, Jr., and D. M. Smith, Aquametry (1948).

47. B. B. Baker and W. N. McNevin, Anal. Chem., 22:364 (1950).

48. GOST 8287-57.

49. GOST 7822-55.

50. N. G. Gaylord, Reduction with Complex Metal Hydrides, New York, Wiley (1956).

51. F. G. Carpenter, Anal. Chem., 34:66 (1962).

52. P. G. Jeffery and P. J. Kipping, Analyst, 87:379 (1962).

53. P. G. Jeffery and P. J. Kipping, Analyst, 87:594 (1962).

54. J. Juranek and A. Ambrova, Coll. Czechosl. Chem. Commun., 25:2814 (1960).

55. J. Juranek, Coll. Czechosl. Chem. Commun., 24:135 (1959).

56. J. M. Walker and G. W. Kuo, Anal. Chem., 35:2017 (1963).

57. W. K. Stuckey and J. M. Walker, Anal. Chem., 35:2015 (1963).

58. T. G. Mungall and D. J. Johnson, Anal. Chem., 36:70 (1964).

59. O. A. Sukhorukov and A. A. Zhukhovitskii, Izv. VUZ, Chern. Met., No. 95, 324 (1964).

60. O. A. Sukhorukov and N. T. Ivanova, Zav. Lab., 31:1078 (1965).

61. F. M. Nelsen and S. Groennings, Anal. Chem., 35:660 (1963).

62. I. Okuno, J. C. Morris, and W. E. Haines, Anal. Chem., 34:1427 (1962).

63. R. L. Bond, W. J. Mullen, and F. J. Pinchin, Chem. Ind., 48:1902 (1963).

64. A. T. Diedrich, R. P. Bult, and J. M. Ramaradhya, J. Gas Chromat., 4:241 (1966).

65. A. D. Zorin, G. G. Devyatykh, V. Ya. Dudorov, and A. M. Amel'chenko, Zh. Neorg. Khim., 9:2226 (1964).

66. K. Porter and D. H. Volman, Anal. Chem., 34:748 (1962).

67. Fang K'ai-ming and Hung Ying-erh, Huahsüeh Shihchieh, 20:75 (1966).

68. L. Dubois, A. Zdrojewski, and J. L. Monkman, J. Air Pollut. Control Ass., 16:135 (1966).

69. G. S. Vizard and A. Wynne, Colliery Engng., 42:353 (1965).

70. W. Schöniger, Mikrochim. Acta, N1:52 (1963).

71. V. N. Lebedeva and N. A. Nikolaeva, Zh. Anal. Khim., 18:984 (1963).

72. I. J. Oita and H. S. Conway, Anal. Chem., 26:600 (1954).

73. K. Tesařik, Gas-Chromatographie, 1965, Vorträge des V Symposiums über Gas-Chromatographie in Berlin, Mai 1965. Nachtrag DAW zu Berlin, p. 89.

74. B. J. Gudzinowicz and W. R. Smith, Anal. Chem., 35:465 (1963).

75. R. L. Bersin, F. J. Brovsaides, and C. O. Hommel, J. Air Pollut. Control Ass., 12:123 (1962).

76. S. Duckworth, D. Levaggi, and J. Lim, J. Air Pollut. Control Ass., 13:429 (1963).

77. E. R. Kulzynski, J. Air Pollut. Control Ass., 13:435 (1963).

78. P. Chovin, J. Lelbe, and H. Mouren, J. Chromat., 6:363 (1961).

79. F. A. Gunther, R. C. Blinn, and D. E. Ott, Anal. Chem., 34:303 (1962).

80. F. H. Huyten and G. W. A. Rijnders, Z. Anal. Chem., 205:244 (1965).

81. F. M. Rappoport and A. A. Il'inskaya, Laboratory Methods of Producing Pure Gases [in Russian], Moscow, Goskhimizdat (1963).

Chapter VII

Accumulation Methods

In general, an impurity analysis consists of the following operations: sample preparation, chromatographic separation, and quantitative determination (detection). If the usual operations are inadequate (e.g., because the detector's response is inadequate or the impurities are not clearly separated from the main component), special preparation methods are used, in particular, accumulation or isolation of the impurities from the base. Also, accumulation (concentration) is used when the impurities are identified by non-chromatographic methods (optical spectroscopy, mass spectroscopy, nuclear magnetic resonance, etc.), and so this step is often a necessary one for complex mixtures.

The treatment increases either the actual impurity concentration

$$O = \frac{c_f}{c_i} \tag{58}$$

(where c_i and c_f are the initial and final impurity concentrations by volume) or the relative impurity concentration

$$\lambda = \frac{\bar{c}_f}{\bar{c}_i}, \tag{59}$$

where $\bar{c}_i$ and $\bar{c}_f$ are the corresponding quantities relative to the main component. Relative accumulation helps to eliminate loss of peaks from overlap by the main peak.

Some methods provide mainly one or other of these forms of impurity enrichment. For instance, adsorption produces higher

actual slow impurity contents, while flushing-out of volatile impurities (with a gas) provides relative accumulation. For instance,
traces of dissolved gases (oxygen, light hydrocarbons, etc.) in organic liquids may be flushed out by bubbling a pure carrier through
the liquid [1-3], and these pass with the carrier to the column for
separation. The carrier has a much higher concentration of these
volatile components than does the liquid, i.e., it is relatively enriched.

In this chapter we consider chromatographic and other enrichment methods.

1. CHROMATOGRAPHIC METHODS OF
IMPURITY ENRICHMENT

These methods may be operated without a carrier gas or as
in elution chromatography.

Preparative Elution Chromatography

This approach is widely used in impurity enrichment because
the separation is highly effective, there is scope for using large
samples, the separation cycle is repeated many times, and the
process is fully automatic.

The method is quite flexible and is applicable to virtually all
volatile compounds. The fractions from the preparative column
are analyzed on ordinary packed or capillary columns or are examined by physical and chemical methods. There are books [4, 5]
and reviews [6, 7] on the theory and practice of preparative gas
chromatography, so here we consider only the preparative isolation of impurities.

Incomplete fraction extraction is a major error source in
quantitative analysis via developmental preparative gas chromatography. This problem is especially important in impurity enrichment because the impurity vapor pressure in the gas may be
very small, which hinders recovery.

Fractions from a preparative column are recovered in cooled
traps, whose design is dependent on the column size, properties of
the impurities, and the subsequent analysis methods [8-21]. A
packed trap appears the most effective, and the packing may be a

chromatographic sorbent. This type of trap is especially convenient when the fractions are analyzed by gas chromatography, e.g.,
in a capillary column. If they are examined by IR spectroscopy,
the packing should be powdered KBr or LiF, which are removed,
pressed, and used directly in spectrum recording.

The preparative stage may be monitored and some possible
errors eliminated by internal-standard methods.

Boggus and Adams [22] were among the first to use preparative elution chromatography in impurity analysis. Figure 31a
shows a typical separation on a preparative column 300×1.25 cm
filled with C-22 celite (40-70 mesh) bearing 20% mineral oil at
40°C with He at 150 ml/min.

The fractions for further analysis began on the tail of the
ethylene chloride peak at 1/32 of the height of the maximum; they
were collected in a trap cooled by liquid nitrogen up to the point
where the slowest component had emerged. The trap was then
connected into the flow system of an analytical column 300×0.625

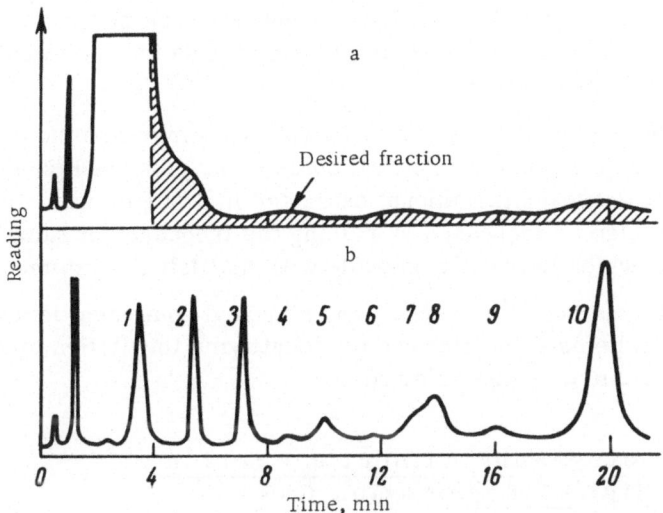

Fig. 31. a) Isolation of impurity fractions on a preparative column;
b) detailed analysis of a fraction [22]: 1) chloroethane, 2) isopentane, 3) n-pentane, 4) bromoethane, 5) 2,2-dimethylbutane, 6)
1,1-dichloroethane, 7) 2-methylpentane, 8) 2,3-dimethylbutane,
9) 3-methylpentane, 10) n-hexane (internal standard, $2 \cdot 10^{-2}$ %).

cm and was heated to release the impurities. This column had the same packing as the preparative column and was operated at 30°C with 50 ml/min of helium. Figure 31b shows the result. The limit of detection was $10^{-6}\%$ when a thermistor katharometer was used. If C_3 compounds have to be determined, the preliminary column should [22] have 20% tricresyl phosphate on C-22, the fractions before the main peak being collected.

Mikkelsen [23] assayed high-boiling impurities in vinyl chloride down to $10^{-6}\%$ in this way. The column received 2 ml of specimen, and about 99% of the main component was discarded at the outlet, all high-boiling components being collected. The impurities up to $5 \cdot 10^{-4}\%$ concentration were identified by mass spectrometer [23], and the same method has been used [24] to identify impurities isolated in this way. See also [25, 26] on this use of the method.

Developmental preparative chromatography is also very effective, but its throughput is not very great. For instance, a laboratory preparative chromatograph might be used (sample 10 ml, separation time 10 min, impurity content $5 \cdot 10^{-4}\%$, 50 mg of impurity needed for identification), in which case 7 days of continuous operation would be needed. Of course, less material might suffice for qualitative analysis, in which case the time would be reduced in proportion.

The column diameter may be raised to increase the throughput [27, 28], or numerous small columns may be used [29-32], in order to increase the amount extracted in a single run. The output can also be increased by raising the frequency of sample injection, which is usually possible with multistage systems [33, 34].

The method is universal, and standard equipment is available, so it can be used in research for identifying impurities by chro matographic and other methods.

Chromatography without a Carrier
Gas (High-Concentration Gas
Chromatography)

The main disadvantage of enrichment via developmental chromatography is that the impurities are eluted at low concentrations, and special methods are needed to extract them from the gas flow.

Zhukhovitskii et al. [35] developed many of the methods of chromatography without a carrier gas, which is also called high-concentration gas chromatography, which provides under isothermal conditions a concentrated output of slow and fast impurities. The working techniques and separation mechanism are different from those of developmental chromatography, since the bands are not separated by layers of carrier gas but are adjacent. Differences in adsorption and gas speed are the causes of bands with sharp boundaries, which show little broadening. The method is capable of yielding an output in the form of the nearly pure components, while differences in flow speed markedly reduce the broadening of the leading and trailing edges [35]. The intermediate step of recovery (by freezing) is not essential, and an output band can be sent directly to an analytical column (for detailed separation into individual components) or to some other instrument (for identification).

There have been several major contributions to the various techniques used in this method [35-43]. Here we consider the principal forms that have been used or might be used for impurity enrichment.

Frontal Chromatography

Here the stream of sample is passed to a column filled with a gas that is indifferent as regards the relevant analysis. Figure 32 shows the output curve for a three-component mixture. The fastest component is obtained in pure form in the first band, so fast impurities (ones sorbed less than the base) can be enriched in this way [44]; some applications have been described [45, 46].

Schay [40] has dealt with the theory and practice of frontal chromatography, so here we consider only topics directly related to enrichment.

One can usually assume that all the fast components have similar Henry coefficients, so the fast zone may be considered as the result of frontal separation of a two-component mixture, in which case the column has the three zones of Fig. 33: I) a mixed zone where the concentrations are as in the initial mixture, II) a fast-component zone, and III) a zone for the gas present before the run.

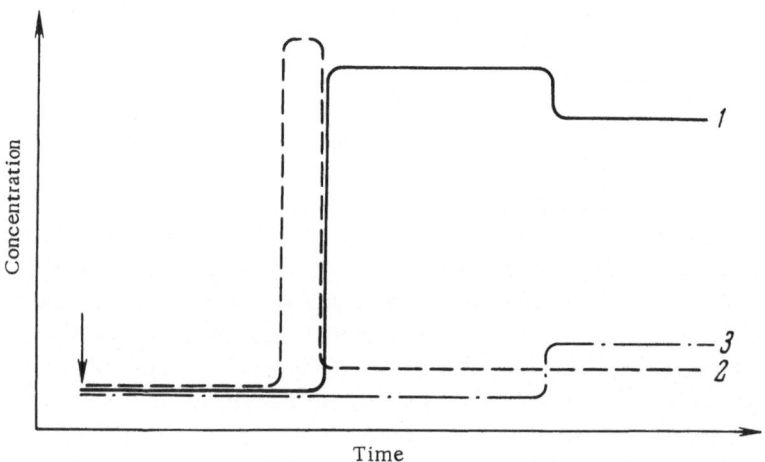

Fig. 32. Hypothetical frontal chromatogram, with concentrations of: 1) main
component, 2) a fast component, 3) a slow component.

We can deduce the speeds of the gas flow and the zone boundaries [46, 47] on the assumption that all the components have linear sorption isotherms. We neglect broadening, and assume that
the concentration c_2^0 of a fast impurity is much less than c_1^0, the
concentration of the main component, i.e., $c_2^0 \ll c_1^0$ and $c_1^0/(c_1^0 + c_2^0) \approx$
1, while the component concentrations in the second and third bands
approach 100%. We also assume that the Henry coefficient* Γ_2 for
the fast component is substantially less than Γ_1 (slow component),
i.e., $\Gamma_1 \gg \Gamma_2$, and also that $\Gamma_2 > \Gamma_3$ (Γ_3 is for the filling gas).

Then the conservation equation for the first major slow component gives us the following relation between the speed u_1 of the

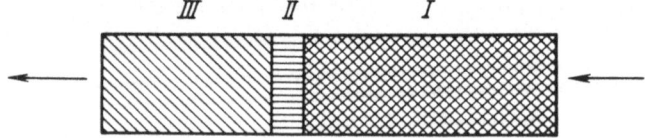

Fig. 33. Frontal separation of a two-component mixture in a
column: I) slow and fast components, II) fast component,
III) fast gas filling column before start of run.

_____

*By Henry coefficient we mean in this book the so-called general Henry coefficient.

boundary of the first band and the linear velocity α_1 of supply of the mixture:

$$\alpha_1 c_1^0 = u_1 \Gamma_1 c_1^0 \qquad (60)$$

or

$$u_1 = \frac{\alpha_1}{\Gamma_1}. \qquad (61)$$

The speed α_2 of the gas flow in the second band is governed by the rate $\alpha_1 c_2^0 - u_1 \Gamma_2 c_2^0$ of supply from the first and by the desorption rate $u_1 \Gamma_2 c_2$ at the boundary between the first and second bands:

$$\alpha_2 = \alpha_1 \frac{c_2^0}{c_2} + u_1 \Gamma_2 \left(1 - \frac{c_2^0}{c_2} \right), \qquad (62)$$

where c_2 is the concentration of the fast component in the second band.

The leading edge of the second band has a speed defined by

$$\alpha_2 c_2 = u_2 \Gamma_2 c_2 \qquad (63)$$

or

$$u_2 = \frac{\alpha_2}{\Gamma_2} = \alpha_1 \left[\frac{c_2^0}{\Gamma_2 c_2} + \frac{1 - c_2^0 / c_2}{\Gamma_1} \right] \approx$$

$$\approx \alpha_1 \left[\frac{c_2^0}{\Gamma_2 c_2} + \frac{1}{\Gamma_1} \right] \approx u_1 \left[1 + \frac{\Gamma_1}{\Gamma_2} \cdot \frac{c_2^0}{c_2} \right]. \qquad (64)$$

The gas speed in the third band is governed by the desorption rate for the filling gas at the boundary between the second and third bands:

$$\alpha_3 = u_2 \Gamma_3. \qquad (65)$$

The impurity-band broadening under these conditions is substantially different from that in ordinary elution chromatography; there are marked gradients in the linear velocities at the boundaries, which result in narrowing. The following equation gives $u_{i,3}$, the velocity ratio of the impurity in the third band and at the front of the second band:

$$\frac{u_{i,3}}{u_2} = \frac{u_2\Gamma_3}{\Gamma_2 u_2} = \frac{\Gamma_3}{\Gamma_2} < 1. \tag{66}$$

Broadening at the trailing edge of the second band (e.g., by diffusion from the second band into the third) is thus restricted because the leading edge of the second band moves faster than does the impurity in the third band.

A similar relation can be derived for the trailing edge of the impurity band:

$$\frac{u_{i,1}}{u_1} = \frac{\Gamma_1}{\Gamma_2} \gg 1, \tag{67}$$

where $u_{i,1}$ is the speed of the impurity along the column in the first band.

It is clear that $u_{i,1} > u_1$, so molecules rapidly return to their own band, i.e., the boundary between bands is sharp.

Then impurity molecules that enter the third band for any reason move at a lower speed, while ones that enter the first band move faster than the boundary. These factors restrict spreading.

Studies have been made [47, 48] of the band width for the fast impurities as a function of the working parameters for $7.5 \cdot 10^{-5}\%$ methane in CO_2 in a 5A molecular-sieve column filled with nitrogen. The emerging methane band was recorded by a flame-ionization detector. This detected only the methane band, because CO_2 is not recorded by that detector. Figure 34 shows the result, where I represents the initial mixture, II is methane, and III is nitrogen. These tests showed that one can simulate enrichment of a largely unadsorbed gas (oxygen, CO, methane, etc.) from olefins (ethylene, propylene, etc.). The method of [47, 48] is convenient for examining the process, as both edges are recorded.

The width μ is related to the linear flow rate α, which is important in choosing the best conditions. Figure 35 shows some experimental results, where $d\mu/d\alpha$ is negative at low α but then becomes positive. The exact μ is dependent on the balance between two sets of opposite factors, one tending to produce broadening

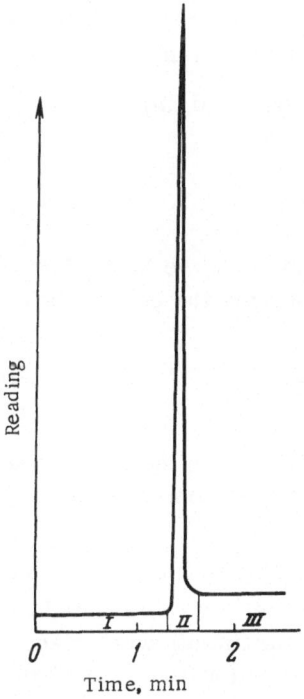

Fig. 34. Frontal enrichment of
methane from CO_2 in a molecular-sieve column 20 × 0.4 cm,
flow rate 50 ml/min.

(eddies, longitudinal diffusion, finite rate of mass transfer, etc.)
and the other tending to produce narrowing (differences in flow
rate in the three zones).

The diffusion controls μ when α is small, so $d\mu/d\alpha$ is negative,
but it becomes positive at high α because the broadening is determined by the finite rate of mass transfer between the phases.

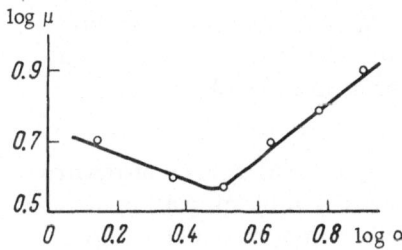

Fig. 35. Width μ of methane band (in
cm^3) as a function of CO_2 linear speed
α cm/sec) in a 40 × 0.3 cm column of
5A molecular sieve.

It is usually desirable to use high α, which substantially reduces the analysis time although it increases μ somewhat.

The amount V_{im} (cm³) of largely unadsorbed impurity in its band is roughly

$$V_{im} = c_2^0 LS (\Gamma_1 - \Gamma_2),\tag{68}$$

where L is column length and S is column cross section. The ratio of V_{im} to the total amount of impurity entering the column is

$$p = \frac{V_{im}}{c_2^0 LS\Gamma_1} = 1 - \frac{\Gamma_2}{\Gamma_1}.\tag{69}$$

In practice, it is important to have p ≈ 1.0, since then the amount recorded by the detector is almost that in the sample taken for analysis.

It follows from (68) that V_{im} increases with the working column volume (length), so we need to examine how μ is related to L at very low concentrations [47, 48]. Measurements were made of μ for methane with L of 20, 40, and 60 cm and a flow rate of 30 ml/min; μ was the same (the peak became higher) in spite of trebling of L. This shows that the broadening in development chromatography (μ increases with L) is essentially different from that in frontal enrichment, where there is only restricted broadening for fast-moving impurities.

The sorbent grain size and the adsorption of the inert gas also influence μ [47, 48], which increases with the grain size and the uptake in sorption. These features must be borne in mind in developing methods.

Frontal enrichment can be used in two ways: with the column initially filled with a gas adsorbed less than is the impurity [44] and with the column evacuated [49]. The first method uses simpler apparatus, but the second is better if a largely unadsorbed gas (Γ_2 small) is to be extracted, when it is difficult to choose a gas whose Γ_3 corresponds to $\Gamma_2 > \Gamma_3$.

Impurities in the initial gas are a potential source of errors [44]. The enrichment band is usually formed by migration of impurity from the base band and by similar impurities given up by

the sorbent (acquired from the filling gas). The proportion of the
latter is

$$\frac{V_{co}}{V_{im}} = \frac{c_{2\,co}^0}{c_2^0} \; \frac{1}{\left(\frac{\Gamma_1}{\Gamma_2}-1\right)}, \tag{70}$$

where V_{co} is the contaminant volume and c_{2co}^0 is the concentration
of the contaminant in the filling gas. Typically, $\Gamma_2/\Gamma_1 \approx 0.01$; if
we assume a permissible error of 1% ($V_{co}/V_{im} \approx 0.01$), then we
must have

$$c_{2\,co}^0 \approx c_2^0, \tag{71}$$

i.e., the contaminant concentration in the gas must roughly equal
the impurity concentration in the sample. This sets a high purity
standard, which is further raised as the sample becomes purer.
Vacuum should be used if no suitably pure filling gas is available.

Frontal enrichment is usually the first step in impurity analysis, the second one being ordinary chromatographic analysis. The
third step is column regeneration.

In the simplest method, the enrichment and analytical columns
are connected in series, i.e., effectively form one column [44, 46].

Frontal enrichment has been used in impurity analysis for
ethylene [44, 46] and propylene [50]. In the first case, the two columns were respectively 80×0.6 cm and 90×0.3 cm, both being
filled with the 0.25-0.5 mm fraction of 5A molecular sieve. The
ethylene sample was 200 ml and the carrier gas was hydrogen.
The response of the katharometer was 800 mV-ml/mg. The temperature of the second column was raised to 57°C when the methane had emerged. Figure 36 shows the analysis of fast-running
impurities in ethylene. The material in the first column was regenerated after each run by heating to 300°C in a stream of hydrogen for 5 min. A single determination required a total time of 25
min. The coefficient of variation was 10%, while the limit of detection was $5 \cdot 10^{-5}$ to $1 \cdot 10^{-4}$%.

A similar method was used for propylene [50], with 13X sieve,
the time for one analysis being 21 min and the limit of detection
$5 \cdot 10^{-5}$%.

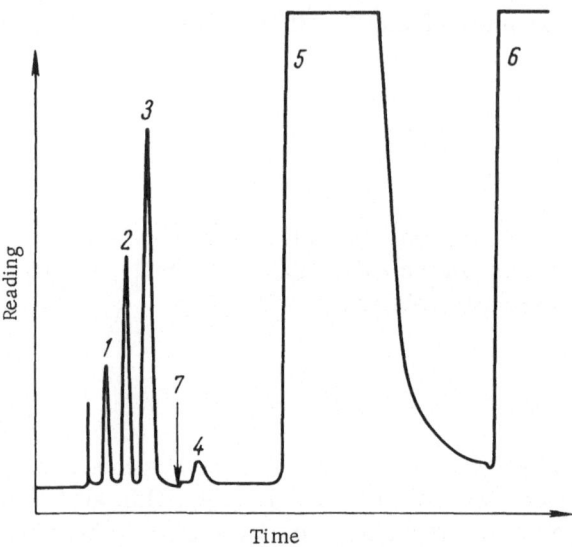

Fig. 36. Analysis of fast-moving impurities in ethylene with
frontal enrichment: 1) oxygen, 2) nitrogen, 3) methane,
4) CO, 5) ethane, 6) ethylene, 7) temperature raised to
57°C.

Frontal enrichment has been performed in the KhTM-IL modification of the standard KhT-63 chromatograph in order to detect
traces of permanent gases in high-purity ethylene [51]. The nitrogen, oxygen, methane, and CO were extracted from a 250-ml sample used with a 330×0.4 cm column of 5A molecular sieve (0.25-
0.5 mm fraction). The analysis was performed with high-purity
helium as carrier. The limit of detection was $5 \cdot 10^{-5}\%$. A calibration mixture (7.5 μl) was injected when the last fast-running
impurity had emerged; in that case, the sample and the standard
mixture were separated on the same part of the column free from
ethylene. The proportions of the impurities in the ethylene were
calculated from the peak heights. The sorbent was regenerated
by a reverse gas flow of 200 ml/min of carrier gas for 20 min,
which is a convenient regeneration method, especially in industrial
use.

Traces of CO in ethylene have been determined [46] by pref-
erential adsorption of the ethylene (150-ml sample) on activated

charcoal (type AR-3, 19 g) in a trap at room temperature, the light impurities (CO, methane, and hydrogen) then being displaced by a flow of air (50 ml/min) at the same temperature. Ethylene and ethane are firmly bound to the charcoal at room temperature, and they were displaced by heating to prepare the material for the next analysis. The limit of detection for CO was $10^{-3}\%$.

The C_4 and C_5 hydrocarbons can [52] be assayed in dimethyl-formamide and acetonitrile by adsorption onto diatomite in a vessel connected immediately before the chromatograph column. A flow of carrier gas took the impurities to the column for analysis, while the solvent remained behind. Two parallel absorbers were used so that regeneration (with nitrogen at 100-150°C for 15-20 min) could be performed during the analysis.

The Dzerzhinsk branch of the Experimental Automatics Designs Office (OKBA) has collaborated with the gas chromatography section of the All-Union Petroleum and Gas Research Institute (VNIGNI) in the development of the Luch laboratory gas chromatograph [52a], which uses frontal enrichment on an evacuated column. The instrument can be used to determine traces of He, Ne, and H_2 in air, hydrogen in argon, etc., at the 10^{-5} to $10^{-6}\%$ level (if 1000-ml samples are used).

Frontal enrichment is a general and effective method now widely used; although published papers relate only to impurities in gases examined with adsorption enrichment, there is no doubt that the method can be used to enrich fast-running impurities from vapor samples on gas—liquid columns.

Frontal chromatography can also be used to enrich slow-running impurities if the desorption (compression) is performed with a hot zone or with a displacing gas [52b, 52c]. In particular, a hot zone is often used in this way. The sorbent in a trap (often cooled) receives perhaps several liters of sample; the main component (less readily adsorbed) passes through and the heavy impurities are selectively retained. Then the trap is connected into the chromatograph system and heated rapidly to displace the impurities into the carrier gas as a narrow band, which is then analyzed in the usual way. This method is widely used [53-58] to detect contaminants in gaseous compounds.

We can estimate the limit of detection in this method if we assume that the main component is not sorbed at all and that we

can neglect the expansion associated with the heating. An impurity can be reliably detected if

$$c_{max} \geqslant 2c_{min},\qquad(72)$$

where c_{max} is the peak concentration after chromatographic separation and c_{min} is the lowest impurity concentration registered by
the detector. Of course, c_{max} increases with sample size, but then
the separation performance tends to deteriorate (Chapter II). The
limiting sample size can be deduced if we assume that c_{max} has
65% of the initial value and there is only 15% loss of performance
[59, 60]:

$$q_{cr} = \frac{1.8\,V_N}{\sqrt{N_0}},\qquad(73)$$

where N_0 is the number of theoretical plates for a small sample.
The impurity can then be detected reliably if

$$c_{max} = 0.65\,c_{des} = 2c_{min},\qquad(74)$$

where c_{des} is the impurity concentration after desorption.

Then after enrichment we should have that

$$c_{des} \geqslant 3c_{min}\qquad(75)$$

with

$$q_{des} = q_{cr}.\qquad(76)$$

The following is the volume of enriched sample going for
analysis

$$q_{des} = q_{in}\ \exp\left[-\frac{Q}{R}\left(\frac{T_{des} - T_{conc}}{T_{des}\,T_{conc}}\right)\right],\qquad(77)$$

where Q is the heat of adsorption, q_{in} is the original sample volume, T_{conc} is the sorption temperature (°K), and T_{des} is the desorption temperature.

From (73) and (77) we have the condition for the sample volume as

$$q_{in} = \frac{1.8 V_N}{\sqrt{N_0}} \exp\left[\frac{Q}{R}\left(\frac{T_{des} - T_{conc}}{T_{des} T_{conc}} \right) \right].$$ (78)

The following is the minimum detectable impurity concentration in the initial sample:

$$c_{in} \exp\left[\frac{Q}{R}\left(\frac{T_{des} - T_{conc}}{T_{des} T_{conc}} \right) \right] \geqslant 3c_{min}$$ (79)

or if the final temperature is high (when $\Gamma_{des} \approx 1$)

$$\frac{c_{in}}{c_{min}} \geqslant \frac{3}{\Gamma_{conc}} \quad \text{or} \quad c_{in} \geqslant \frac{3}{\Gamma_{conc}} c_{min}.$$ (80)

Equations (79) and (80) allow one to estimate the improvement in the detection limit provided by the method if the final sample is not too large.

The length of the trap should not be less than $q_{in}/S\Gamma_{conc}$, where S is the cross section; this is the length needed to retain all the heavy impurities. If Γ is large, enrichment can improve the detection limit by over a factor 100.

The above equations give the basic specification for the process parameters.

The trap is best filled with a high-capacity sorbent, of which activated charcoal is the commonest.

Mosen and Buzzelli [53] determined traces of oxygen, nitrogen, CO, CO_2, and methane in He with a trap (2-mm glass tube) filled with 0.75 cm^3 of activated charcoal that had been heated to 150-200°C under vacuum for an hour. The trap was cooled in liquid nitrogen. After the sample had passed through, the trap was heated rapidly to 90°C, and the desorbed compounds were carried by a flow of He to the column of 5A molecular sieve (analysis temperature 50°C, flow rate 34 ml/min). The CO_2 was assayed on a silica gel column at 70°C. The limits of detection were 10^{-2} to $10^{-4}\%$.

Various sorbents have been tested [58] by using a trap receiving 5 μl of benzene in 20 liters of air followed by desorption
at 200°C. The peak area was compared with that on injecting the
same amount of benzene directly into the analysis column. Activated charcoal (0.2 g) gave the best results, but not all compounds
were recovered fully. With adsorption at 0°C and release at 200°C,
the recovery percentages were benzene 85, acetonitrile 100, nitromethane 91, methanol 83, ethanol 90, and acetone 80.

Low-temperature adsorption on charcoal has been used [58a]
in assaying traces of hydrogen in gases.

Molecular sieves are very useful in enrichment. Brenner et
al. [61] examined the selectivity of 5A molecular sieve for compounds and classes of compounds and found that the following were
retained: n-alkanes, iso-alkanes, aromatic compounds, n-olefins,
cycloalkanes, isoolefins, n-alcohols, aldehydes, acids, esters, and
halides. Compounds not retained were iso-alcohols, ethers, inert
gases, carbon monoxide, oxygen, nitrogen, methane, nitromethane,
dimethyl sulfide, thiophen, and carbon disulfide. The sieve can
thus be used to enrich compounds of the first group present as
traces in compounds of the second group.

Various grades of silica gel are also used in enrichment;
Brenner and Ettre [57] used adsorption on silica gel to assay impurities in various gases (Fig. 37). The 50 × 6 mm trap was
cooled to − 80°C in acetone + dry ice. A known volume of sample
was passed with the stopcock in position 1, and then (with the trap
still cooled) a flow of helium was passed with the stopcock in position 2 to remove traces of the major component. The end of flushing was registered with the chromatograph recorder (sometimes
it was sufficient to connect the trap to a vacuum line in order to
remove the major component). Then the stopcock was returned
to position 1 and the trap was placed in a bath at 50-60°C to release the adsorbed gases rapidly. After thermal equilibration, the
trap was connected to the chromatograph (position 2) for analysis.
The system was used in various applications; Table 10 gives the
basic characteristics for some of these.

Yavorovskaya [56] used silica gel to enrich chloromethanes,
C_n tetrachloroalkanes (n odd) up to C_6, and chlorobenzenes in the
presence of benzene in air. Water vapor did not interfere with the

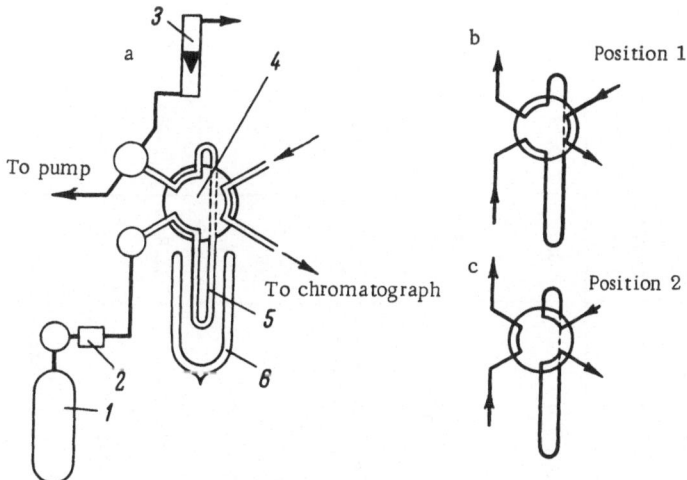

Fig. 37. System for enriching heavy impurities in gases: a) general system, b) trapping, c) desorption; 1) sample source, 2) flow regulator, 3) rotameter, 4) gas stopcock, 5) trap, 6) dewar for coolant.

determination because nephelometry and colorimetry were used as the methods of analysis. The sampling time was reduced by passing the air through a fluidized bed of silica gel at 5-6 liters/min. The temperature was raised to release the impurities. The recovery was found to be somewhat less than when the same amount of material was injected directly into a chromatography column.

Use should be made of experience [62-65] with microtraps in preparative gas chromatography if one wishes to extract heavy compounds from largely unadsorbed gases at low temperatures. Traps are commonly filled with sorbents for gas—liquid chromatography, e.g., the trap of [2] was a 9×0.3 cm tube filled with 60-80 mesh celite bearing 7% Apiezon. The trap could then be heated to 220°C to relase the compounds.

Eggersten and Nelsen [66] assayed the C_2-C_6 hydrocarbons in air with a column 30×0.8 cm filled with 20-30 mesh diatomite bearing 40% dimethylsulfolane; the 5-10 liter sample was passed at -183°C after drying and removal of acid gases in an Ascarite column. The trap was then flushed with He for 20-30 min while con-

TABLE 10. Use of Enrichment in Heavy-Impurity Determination by Gas Chromatography

Main compound	Impurity	Trap	Column	Sample volume, liters	Limit of detection $10^{-4}\%$
Air	CO_2	Silica gel, adsorption at $-80°C$, desorption at $+50°C$	Silica gel, 2 m	0.5	0.17
Air (oxygen)	C_2H_2	Silica gel, adsorption at $-80°C$, desorption at $+50°C$	Silica gel, 2m, 65 ml/min, 50°C	8.9	0.13
Hydrogen	N_2, O_2, CH_4, CO	Silica gel, adsorption at $-80°C$, desorption at $+50°C$	Molecular sieve 5A, 2m, 80 ml/min, 50°C	1	1-5
Hydrogen	CH_4, C_2H_6, CO_2, C_3 hydrocarbons, C_2H_4	Silica gel, adsorption at $-80°C$, desorption at $+50°C$	Silica gel, 2m, 80 ml/min, 50°C	16	0.1-0.3
Hydrogen	Higher hydrocarbons	Polyethylene glycol on Chromosorb, adsorption at $-80°C$, desorption at $+50°C$	Dimethylsulfolane on Chromosorb, 4m, 75 ml/min, 50°C	12.8	1-2
Ethylene	C_2H_2	Polyethylene glycol on Chromosorb, adsorption at $-80°C$, desorption at $+25°C$	Dimethylsulfolane on Chromosorb, 4m, 90 ml/min, 25°C	40	1

nected to the chromatograph, and then it was quickly heated to 0°C, the compounds being carried by the He through another Ascarite column to a 750×0.6 cm column having the same packing as the trap. The quantities were deduced from calibration curves; Fig. 38 shows the result for impurities in city air. The limit of detection was $10^{-5}\%$ with a thermal-conductivity detector.

Williams [67] devised important enrichment methods for determining atmospheric contaminants. The air was drawn at 0.5 liter/min for 5 min through a potash column (drying agent) or a magnesium perchlorate one, and then through a 20.3×0.6 cm trap filled with Chromosorb P treated with dibutyl phthalate at −80°C. The trap was then evacuated to remove residual air and was heated to 120°C. The impurities were identified by the use of selective detectors and reactive gas chromatography.

Methane at the $10^{-4}\%$ (by volume) level has been assayed in air [54]. The 60×0.6 cm trap had 15% dinoxyl phthalate on activated charcoal, which was cooled in dry ice; the sample volume was 2.8 liters. The compounds were desorbed at 350°C and the vapors were transferred to a known volume with a mercury pump for gas chromatography. The coefficient of variation was 7.9% at the $30 \cdot 10^{-4}\%$ level.

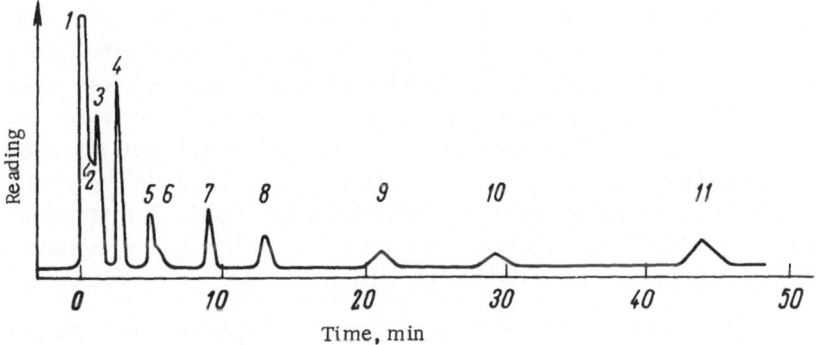

Fig. 38. Impurities in air after enrichment [66]: 1) air, 2) ethane, 3) ethylene, 4) propane + N_2O, 5) propylene, 6) isobutane, 7) n-butane, 8) acetylene, 9) isopentane, 10) n-pentane, 11) propyne + pent-1-ene.

Carlstrom et al. [68] used the selective retention of water by polyethylene glycol to assay water in butane. The 30×0.6 cm trap had 30% of grade 200 glycol on 20-30 mesh diatomite and was kept at 10°C during passage of the sample. A flow rate of 100 ml/min could be maintained for 4 hr before the water broke through. Desorption at 90°C was performed with helium flowing in the reverse direction, which required a smaller volume of helium. The water was assayed with a 60-cm column having the same packing, the gas flowing at 100 ml/min. The retention time of the water under these conditions was 11.5 min. A 10-liter sample gave a limit of detection of $0.2 \cdot 10^{-4}\%$. The method undoubtedly could be used for water in other gases.

A similar method was described in [69]. The trap 0.5 m long was filled with 20 wt.% grade 150 polyethylene glycol on Fluoropak, with adsorption at 0°C and desorption at 90-100°C. The column was 1 m long, had the same filling, and was operated at 90°C with 100 ml/min of helium. The limit of detection was $9.5 \cdot 10^{-3}\%$ for a 560-ml sample.

Cropper and Kaminsky [58] assayed toxic organic compounds in air by trapping in tubes 6.5 cm long and 4.5 mm in diameter filled with various materials. The tube was then connected to a chromatograph (Fig. 39) and was heated rapidly to pass the impurities to a gas–liquid column.

A special study has been made [70] of enrichment of compounds present in air in very small amounts, to detect biochemical differences between individuals, and to locate faults in equipment. It was found that freezing and adsorption in such cases often gave incomplete extraction. The trap was a spiral of pyrex glass whose walls bore a layer of liquid. The trapped compounds were released by heating and were reabsorbed in a trap cooled in liquid nitrogen and containing a molecular sieve. This two-stage enrichment gave access to concentrations below 10^{10} molecules/cm^3. A method has been described [71] for detecting impurities below $10^{-7}\%$ in helium by enrichment in a trap cooled in helium.

High desorption temperatures can produce undesirable reactions, especially if the compounds are unstable, and so sometimes it is better to use extraction with a suitable solvent.

Franst and Hermann [72] examined the effects of working conditions and trap parameters on the extraction of atmospheric con-

Carrier gas

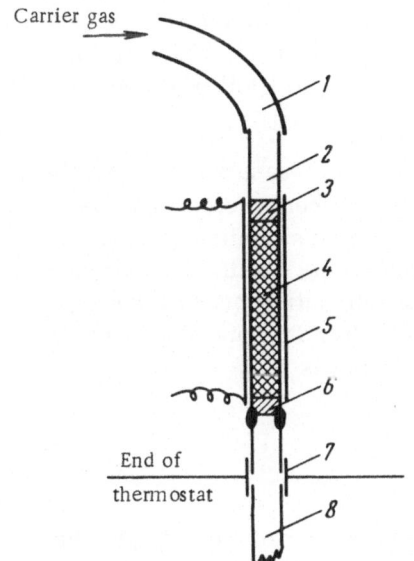

End of

thermostat

Fig. 39. Connection of a trap to a column in impurity desorption [58]: 1) silicone rubber, 2) adsorption tube, 3) glass wool, 4) adsorbent on a carrier, 5) heater, 6) glass wool, 7) silicone rubber, 8) chromatographic column.

taminants onto activated charcoal. The 8.3×0.6 cm column contained 0.7 g of charcoal and received air with known concentrations of the substances. After saturation, the charcoal was washed for an hour with 10 ml of CS_2, and 5 μl of the solution was analyzed. The resulting concentrations were compared with the initial ones. Large grain sizes produced less efficient trapping, but the air volume had no effect. The best flow rate was about 100 ml/min, but the value should be determined by trial in each particular case.

The C_1-C_4 alcohols in air have been assayed [73] by a displacement form of extraction from KSS-3 silica gel (particle size 0.088-0.149 mm) in a glass column 15×0.2 cm, which received 0.5 liter of air followed by impurity displacement by water, which was chosen because it is adsorbed more strongly than any of the alcohols by silica gel. Also, a flame-ionization detector is largely insensitive to water, so there is no need to consider separating the water from the alcohols. Displacement has the further advantage that the impurities are eluted from the gel as a narrow band ahead of the water.

It was found [73] that almost all of the adsorbed alcohols were eluted in a volume of 0.10-0.12 ml with not less than 90% recovery.

The analysis was done on a Spherochrome-1 column with 10% cetyl alcohol 3.5 m long and 4 mm in diameter at 90°C, with nitrogen at 45 ml/min. Table 11 gives the results. The limit of detection was 3 mg/m^3.

All of the above methods involve complete uptake of readily adsorbed compounds. An essentially different method was described by Novak et al. [74], where the gas sample was passed through a short extractor column containing a suitable stationary phase at room temperature until the impurities broke through, i.e., the sorbent was in equilibrium with the impurities throughout its length. The total amount in the trap was then

$$q = LSKc_0, \tag{81}$$

in which K is the partition coefficient at the working temperature for unit volume. As K and q are dependent on temperature, the trap has to be kept at a constant temperature. The sample volume may equal or exceed the value needed for equilibration by the impurity of highest K. The minimum necessary volume is

$$V_{\min} = V_R \left[1 + \sqrt{\frac{8 \ln C\alpha}{L}} \right], \tag{82}$$

where V_R is the retention volume for a column having the size and packing of the trap, C is a constant that characterizes the mass transfer, α is the linear velocity of the gas, and L is the packed length.

The trap was a tube 4.5 cm long and 0.5 cm in diameter, with one end open and the other end joined to a syringe needle. The packing was [74] usually celite-545 bearing 30 wt.% E-301 silicone elastomer. The open end was joined to a vacuum pump and the other to the gas source. After the necessary sample volume had been passed, the tube was connected to a gas chromatograph. Slight variations in gas pressure and flow rate do not affect the baseline stability of a flame-ionization detector. Figure 40 shows curves for impurities in air with: a) direct analysis of 10 ml, b) preliminary enrichment on 0.3 g of sorbent containing E-301 at 24°C.

TABLE 11. Determination of Alcohols in Air by Adsorption Followed by Displacement

Alcohol	Analysis 1			Analysis 2			Analysis 3			Permitted limits,* mg/m³
	S_x, mm²	$q_x \cdot 10^{-4}$, mg	Initial concentration, mg/m³	S_x, mm²	$q_x \cdot 10^{-4}$, mg	Initial concentration, mg/m³	S_x, mm²	$q_x \cdot 10^{-4}$, mg	Initial concentration, mg/m³	
Methanol	552	21.3	4,26	219	17.1	3,41	374	19,8	3,96	50
Ethanol	726	15.3	3,06	406	17.3	3,46	545	15,9	3,18	1000
Isopropanol	782	13.6	2,72	451	14.9	2,98	14,7	14,7	2,94	—
t-Butanol	920	10.4	2,08	525	11.1	2,22	704	10,8	2,16	—
n-Propanol	890	13.8	2,76	462	14.5	2,90	685	14,3	2,86	200
sec-Butanol	1160	14.0	2,80	597	14.7	2,94	862	14,5	2,90	—
iso-Butanol	2240	23.0	4,60	1090	22.5	4,50	1600	22,5	4,50	—
n-Butanol	1460	16.6	3,32	790	20.0	4,00	1160	20,0	4,00	200
n-Propanol (standard)	500	7.0		247	7.0		362	7.0		
Total content† (under normal conditions), mg/m³		28.8			29.6			29.6		

*No limits have been laid down for the iso-alcohols.
†Chemical assay in all cases gave 30 mg/m³.

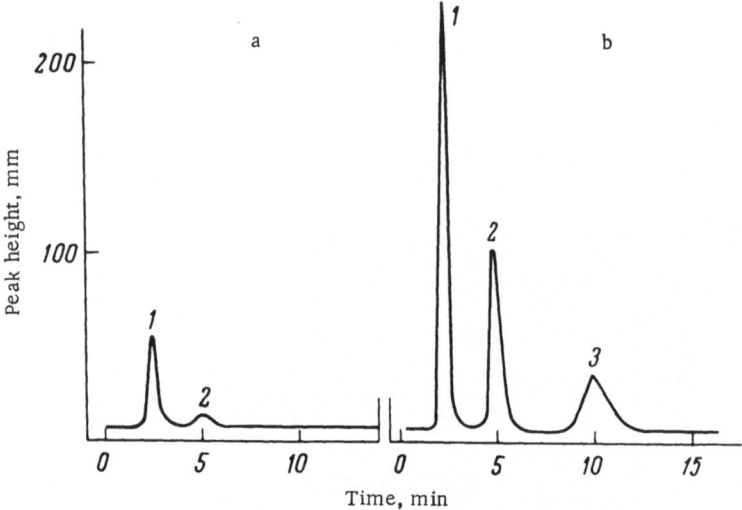

Fig. 40. Assay of organic vapors in air by saturation of an adsorbent [74]: a) direct analysis of a 10-ml sample, b) analysis after enrichment on 0.3 g of sorbent; 1) benzene, 2) toluene, 3) xylene.

This method has the following advantages from the analytical viewpoint:

1. There is no need to monitor the exact volume of sample passed through the trap; it is only necessary to use an excess and to know the trap temperature exactly;
2. It is possible to improve the limit of detection selectively or to remove interfering components, by proper choice of packing; e.g., a nonpolar liquid can eliminate the effects of water vapor, which is the main source of difficulty in sorption enrichment;
3. The trap can tend to equalize the amounts of the individual components, because the partition coefficient increases with the molecular weight and as the vapor pressure decreases.

This last effect is clear from Fig. 40. A slight disadvantage is the need to keep the trap temperature constant.

A form of the frontal method has been used [75-77] in the analysis of low-volatility compounds, and the method has found many applications [78-83]. In this method, an oven with a reverse temperature gradient is passed periodically along the column, which receives the sample mixture continuously. The slow-moving components adsorbed at the start of the column are displaced along the column by the oven. The main component of the mixture acts as the carrier in this case.

The slow components may be separated into bands under the conditions of steady-state chromatography at the same time as they are compressed. The speed of the oven must be considerably greater than that produced in the bands by the gas flow, of course.

Figure 41 illustrates the band production. See [4] on the theory. A decisive parameter is η, the ratio of the oven speed to the

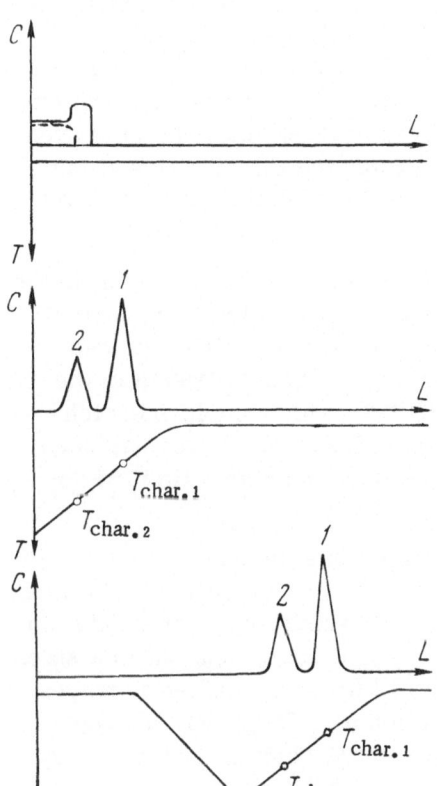

Fig. 41. Production of a slow-component band with a moving oven, where c is the concentration of the first component (solid line) or the second one (broken line), T is temperature, and L is column length.

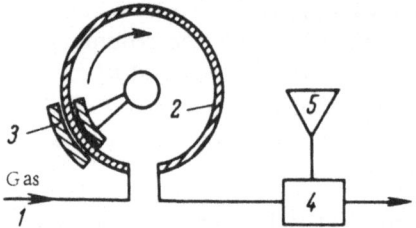

Fig. 42. Apparatus for enriching slow components in a ring column: 1) mixture, 2) column, 3) moving oven, 4) detector, 5) recorder.

speed of the mixture, and the enrichment increases as η decreases [4]. For instance, an enrichment factor of 25 was obtained with an initial butane concentration of $6 \cdot 10^{-4}\%$ and $\eta = 0.4$, while $\eta = 4.5 \cdot 10^{-3}$ gave a factor of 117. The process takes about 12 min on a 100-cm column.

Figure 42 shows the design of a useful ring column that receives the mixture continuously. The oven also moves continuously. The slow components accumulate at the cool start of the column when the oven is at the end, and these are released and separated when the oven moves to the start. The separated slow components leave the column as the oven reaches the end. This oven method requires no dispensing unit, and the average slow-component concentrations can be determined automatically at set intervals.

The method is generally useful in gas borehole logging, in the analysis of inert gases, and in the analysis of slow components generally. The Kazakh Petroleum Gas Research Institute has collaborated with the Institute of Oxygen Machine Design in the design of the KhTD chromatograph [78], which employs enrichment and separation in a sample-gas flow. The device has been used in the analysis of krypton concentrates, with a limit of detection of $4 \cdot 10^{-3}\%$ for krypton.

Vagin and Petukhov [79] have used this method extensively in the production of oxygen and inert gases in a modification in which the slow components are displaced and separated after frontal enrichment not in a stream of the principal component but in a stream of a special carrier gas. The method has been applied to traces of acetylene in oxygen (limit of detection $5 \cdot 10^{-7}\%$), CO_2 in oxygen ($2.5 \cdot 10^{-5}\%$), N_2 and O_2 in argon ($5 \cdot 10^{-3}\%$), and Kr and Xe in oxygen ($5 \cdot 10^{-3}\%$).

If it is difficult to separate the slow components, the method is best used only to enrich them, the separation being performed by elution under isothermal conditions. Petukhov and Vagin [80] have used this approach to determine traces of heavy hydrocarbons in krypton concentrates with a limit of detection of 10^{-4} to $10^{-5}\%$. Kaiser [83a] showed that the method can give access to impurities at the $5 \cdot 10^{-10}\%$ level.

The column length restricts the amount (concentration) of the material in this method. A circulation system can be used [84] to increase the length of the layer. If two oven systems are used together (Fig. 43), the concentration of the enriched components can be greatly increased. In situation I, the gas passes through the first column, the detector, and the second column before escaping.

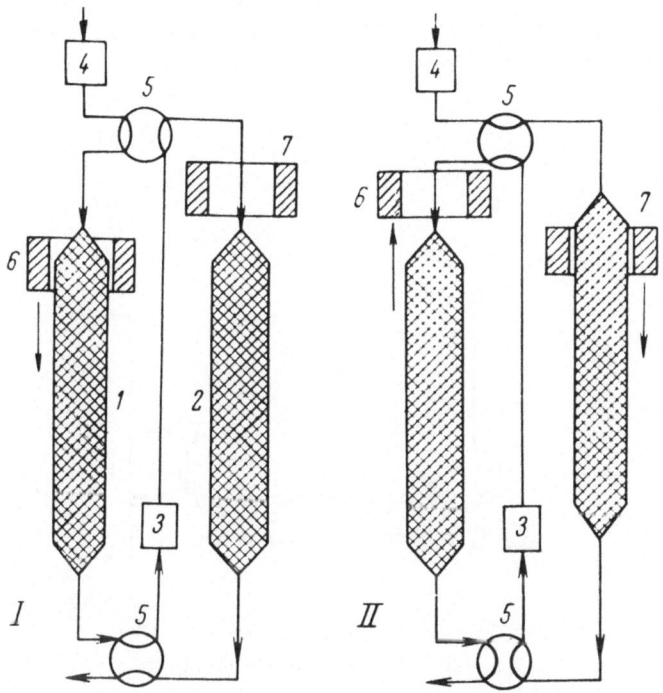

Fig. 43. Circulation system for enriching slow components: 1) and 2) columns, 3) detector, 4) dryer, 5) stopcocks, 6) and 7) heaters (ovens).

When the oven reaches the end of the first column, the slow components pass from the first column into the second. The oven on the first column returns rapidly to the initial position and the flow pattern is changed to that of II, where the gas passes first through the second column and then through the first one.

The frontal enrichment is followed by compression of the bands by the moving oven on the second column. The amount in the moving band equals the sum of (1) the amounts extracted from the mixture entering the second column and (2) the amounts entering the second column after enrichment in the first. Then the band passes to the first column and the cycle is repeated. The effective column length and impurity concentration increase with the number of cycles. For instance, such a system raised the impurity concentration from 0.22 to 80% [84].

Kirkland [85] used desorption from an enrichment column by a moving oven to assay high-boiling herbicides in industrial products (Fig. 44). Up to 10 ml of the sample can be injected into the evaporator of the enrichment column, whose temperature is some-

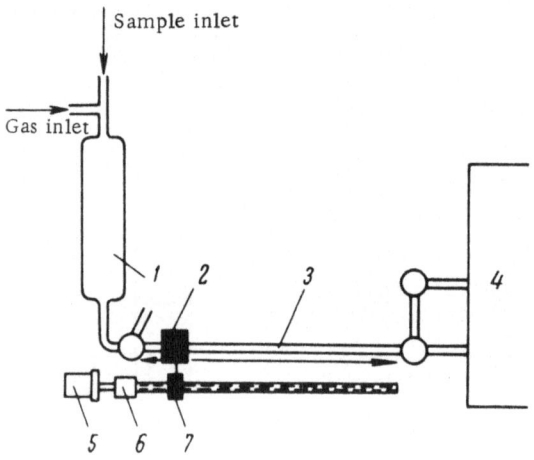

Fig. 44. Enrichment column with moving oven for extracting high-boiling components for injection into a chromatographic column [85]: 1) evaporator, 2) oven, 3) trap, 4) chromatograph, 5) electric motor, 6) speed regulator, 7) oven drive.

what above the boiling point of the solvent. All of the solvent is vented to the atmosphere, while the high-boiling impurities remain in the column and are displaced by a moving oven. This is an alternative form of the method and serves to produce a narrow band for the target compounds at the chromatograph inlet. Also, the impurities may be displaced fractionally by gradually raising the oven temperature.

The chromathermographic method has been used in impurity enrichment [86]. The mixture passes through a column (trap) having a sorbent cooled on one side and heated on the other. The sorbed impurities migrate to the cold end in response to the temperature gradient, and this part of the material is isolated from the main body for desorption into a carrier gas or extraction into a small volume of solvent.

Enrichment onto a moving sorbent [87] is another form of the method, which resembles the circulation method in giving high degrees of enrichment with a short column (a 9-cm working length was used [87]). This method has the sorbent moving against the flow, and the linear velocity of the impurity band should be less than that of the sorbent. If there is a hot region at the inlet, the slow impurities should accumulate at the boundary between the hot and cold parts. Figure 45 shows the apparatus of [87]. The carrier gas or sample enters the glass column 8 containing the moving sorbent via a four-way stopcock; the column was 90 mm long and had an internal diameter of 4 mm. The 0.4-0.6-mm fraction of ASK silica gel moves downwards under gravity in response to the vibration at about 150 mm/min, which is governed by the hole size in the stop. The analysis column had 7% squalane on Sterhamol, length 1 m. The sample gas was dispensed exactly by using a capillary whose resistance for the carrier gas was close to that of the gas line. Two rheometers were used to set up equal flow rates of carrier gas through the apparatus and of sample through the capillary. The sorbent was set in motion, and then the four-way stopcock was turned to pass a set sample volume through the moving-sorbent column, after which the supply of sorbent was stopped. The sorbent was passed through the hot zone to release the accumulated slow impurities. The method was used to extract 0.001% butane from helium with an enrichment factor of about 100.

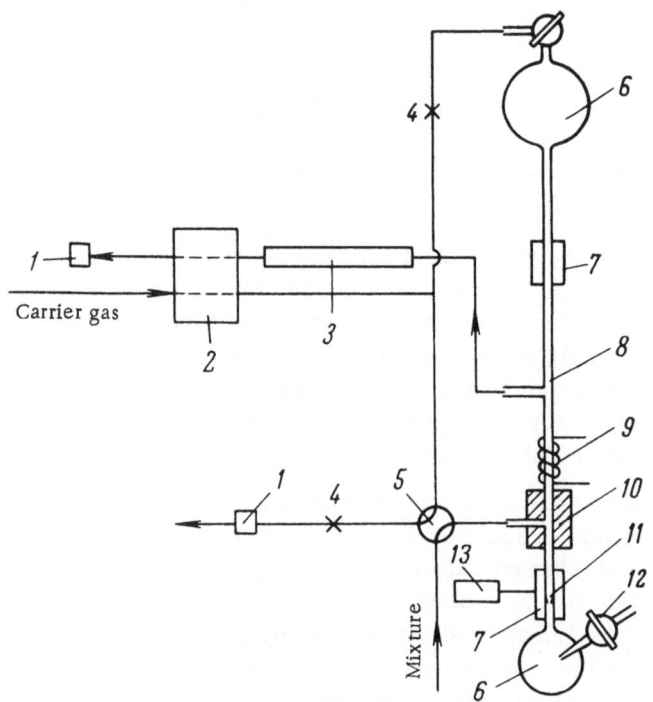

Fig. 45. System with a moving sorbent for enriching slow-moving impurities: 1) rheometer, 2) katharometer, 3) separating column, 4) capillary, 5) four-way stopcock, 6) vessel for sorbent, 7) rubber tube, 8) enrichment column, 9) water-cooling coil, 10) heater, 11) stop, 12) stopcock, 13) vibrator.

Frontal chromatography is thus effective in enriching slow and fast impurities.

Chromatography by Displacement, Thermal Displacement, and Elution Combined with Thermal Displacement

Apart from frontal techniques, impurity concentration can be performed by thermal displacement, simple displacement, and elution combined with thermal displacement.

The first studies on thermal displacement are to be found in
[88, 89], while studies of the mechanism are to be found in [90-92].
The sample (which may contain slow and fast components) is fed
to a column until the sorbent layer is saturated; then an oven is
moved along from the inlet end. Repeated sorption and desorption
result in nearly complete separation of the components. In the
steady state, adjacent bands for pure components move along with
the speed of the oven: the fast-running ones as the first band and
the slow-running ones nearer the oven (Fig. 46). The flow of slow
components released by the heat serves to displace the faster components, which is the origin of the pure bands. Broadening is reduced if the differences in the adsorption behavior of the components is pronounced. Diffusion and the oven speed also affect the
result. This method has been used in the preparation of neon [93],
krypton [94], and propylene [95-97], and also for preliminary enrichment of impurities in propylene [95].

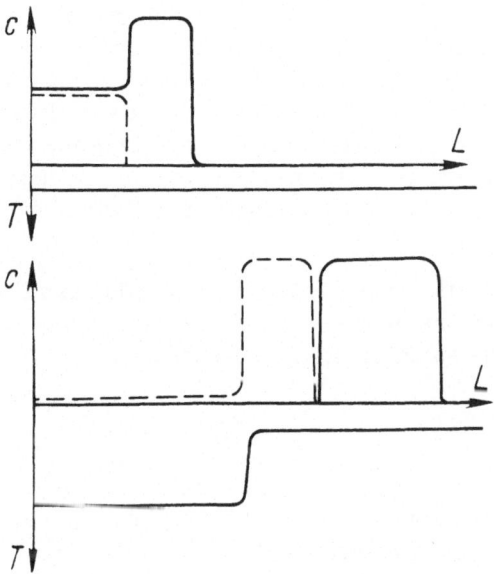

Fig. 46. Band production in thermal displacement
chromatography. See Fig. 41 for explanation.

Thermal displacement can be used to enrich fast and slow components for subsequent detailed analysis by chromatographic or other methods.

It should be borne in mind in enriching slow components that often these lie in the hot region, which can result in reaction or decomposition, which can alter the compositions of the slow and fast fractions. This disadvantage is especially pronounced if the desorption temperature approaches the decomposition point under the working conditions. For instance, it has been shown [98] that the method cannot be used to prepare compounds such as CCl_4 or diethyl ether on wide-pore silica gel because of decomposition during separation under the conditions used.

This disadvantage can be eliminated if the compounds are displaced by a readily adsorbed compound (gas piston) instead of by heat. The compound should be one that is released on heating without decomposition during regeneration at an elevated temperature. If the compound also produces no response in the detector, the scope of the method is greatly extended [97a].

Glückauf and Kitt [99] and Roginskii et al. [100, 101] have described preparative displacement chromatography (a form of chromatography without a carrier gas). The bands are produced as discussed above for thermal displacement, except that a gas piston is used instead of heat. The column is filled to the extent of 50-60% by treatment with the initial mixture, and then the piston gas is admitted. The separated components occur in the sequence of the reciprocals of the retention volumes, and their concentrations range up to 100%.

The speed differences between the bands result in separation under the conditions of gas—liquid chromatography. The speed of the piston in the steady state is defined by

$$u_{pist} = \frac{\alpha_{pist}}{\Gamma_{pist}}, \tag{83}$$

where α_{pist} is the linear velocity of the piston gas and Γ_{pist} is the corresponding Henry coefficient.

The gas speed in the slow band is governed by the desorption rate at the boundary with the piston. The desorbed material travels to the boundary of the band against a more mobile component:

$$\alpha_s = u_{pist}\Gamma_s . \tag{84}$$

The leading edge of the slow-component band moves with the speed of the piston:

$$u_T = \frac{\alpha_s}{\Gamma_s} = \frac{u_{pist}\Gamma_s}{\Gamma_s} = u_{pist} . \tag{85}$$

The linear velocity in the fast-component band is given by

$$\alpha_f = u_{pist}\Gamma_f , \tag{86}$$

where Γ_f is the corresponding Henry coefficient.

There are marked velocity gradients at the boundaries between bands, and these restrict the broadening. As an example we consider the production of the slow band.

The speed u_{sp} of the slow-component molecules along the column in the piston band is higher than u_s, the speed in the slow band:

$$\frac{u_{sp}}{u_s} = \frac{\alpha_{pist}/\Gamma_s}{\alpha_s/\Gamma_s} = \frac{\alpha_{pist}}{\alpha_s} = \frac{u_{pist}\Gamma_{pist}}{u_{pist}\Gamma_s} = \frac{\Gamma_{pist}}{\Gamma_s}, \tag{87}$$

and so $u_{sp} \gg u_s$, since $\Gamma_{pist} > \Gamma_s$ from the condition for displacement. This means that slow-component molecules that enter the piston band for any reason rapidly return to the slow band. Similar relationships apply to the leading edge of the slow band, because u_{fs} (slow-component speed in the fast band) will be substantially less than the speed of the leading edge of the slow band:

$$\frac{u_{fs}}{u_s} = \frac{\alpha_f/\Gamma_s}{u_{pist}} = \frac{u_{pist}\Gamma_f}{u_{pist}\Gamma_s} = \frac{\Gamma_f}{\Gamma_s} \ll 1, \tag{88}$$

since $\Gamma_f < \Gamma_s$. Any molecules of the slow component that get ahead of the band are thus soon caught up by the slow band.

This sharpening mechanism in carrier-free chromatography does not apply in gradient-free developmental chromatography.

The effect increases with the differences between the Γ for the components, and the method is ineffective for compounds with similar Γ. Unfortunately, no study has been made of the effects of the ratios of the Γ on the separation. Displacement is best applied to enrichment of fast and slow impurities whose Γ are substantially different from Γ for the main component.

Some restriction is implied by the need to use a very pure piston compound, because any impurities accumulate along with those from the sample. Thermal displacement does not have this disadvantage because no piston compound is used. The main disadvantage of thermal displacement (decomposition of slow components in the hot region) is overcome in thermal displacement with elution [102, 103], where the separation is produced by a moving oven in conjunction with a slow flow of carrier gas. The sample is fed to a column that has been flushed with inert carrier until the initial section of the column is saturated; then the oven is set moving and the carrier gas is set flowing slowly. The oven produces considerable desorption in the hot region, so even a slow gas flow is quite sufficient to shift a component rapidly to a cooler part of the sorbent ahead of the oven.

The uptake increases considerably when the cold part is reached, and hence the speed falls. The oven continuously displaces the component into the gas, which carries it forward, in this way. The moving slow-component zone impels the fast-component zone by displacement, so the conditions ahead of the oven resemble those without a carrier gas, except that here there is a small gas flow, though this cannot have a major effect on zone production and movement.

This method can take two forms: (1) the gas speed exceeds the oven speed (which appears to be preferable), (2) the two speeds are equal. The process occurs if

$$u_{hot} > w_o > u_{col},\tag{89}$$

where u_{hot} is the speed of the readily sorbed component in response to the gas in the hot zone, u_{col} is the same in the cold zone, and w_o is the speed of the oven. Also,

$$T_d > T_x < T_{max},\tag{90}$$

where T_x is the temperature at which $u_{hot} = w_0$, T_{max} is the maximum temperature in the oven, and T_α is the temperature for onset of decomposition. Further, T_x is given by

$$w_0 = u(T_x) = \frac{\alpha}{\Gamma_s(T_x)}, \tag{91}$$

where $u(T_x)$ is the slow-component speed in the hot zone, α is the linear velocity of the gas, $\Gamma_0(T_x) = A \exp(Q/RT_x)$ is the Henry coefficient for the slow component at T_x, and Q is the heat of sorption.

In the steady state, ahead of the oven are closely adjacent bands for the pure components, whose edges move at w_0, and so the separation conditions are similar to those without a carrier gas, except that here there is a small gas flow, though this cannot have a major effect on zone production and movement. At the boundaries there are gradients in the linear velocities, which restrict zone broadening.

Broadening has been examined [102, 103] in relation to α, w_0, and sample composition, but these have little effect under most conditions.

An example is impurity enrichment from chemically pure hexane by elution and elution with thermal displacement (1.0-ml sample in both cases) [103]. The second method gave enrichment factors of 180–400 for some compounds, which provided detection of an impurity remaining undetected in the elution method. This illustrates the advantages of the second method.

Combined Use of Methods from Liquid and Gas Chromatography

Liquid chromatography is valuable for enriching impurities in organic compounds because it is generally more selective than gas chromatography in separating high-boiling compounds. One can use the method in column, thin-layer, and other forms.

For instance, traces of cyclohexanol have been determined in 50 ml of toluene [104] by passage through a silica gel column; then the toluene was displaced from the free space with air and the adsorbed cyclohexanol was eluted with 2 ml of diethyl ether. This produced an enrichment factor of 25. The limit of detection was $6 \cdot 10^{-4}\%$.

Liquid chromatography can be used in group identification of impurities, e.g., nonpolar trace components in ethereal oil [105] previously freed from all polar components by liquid chromatography on 100-200 mesh silica gel (40 g), with 75 g of sample. This retained the polar compounds strongly.

Hydrocarbons in beeswax have been assayed similarly [106]. The sample was taken up in petroleum ether, which was passed to a 30×1.6 cm alumina column, which was then eluted with petroleum ether. The output fractions were examined by gas—liquid chromatography on Chromosorb bearing 20% of diethylene glycol succinate at 225°C, carrier gas He.

Scott and Phillips [107] assayed heptenes in heptane by passing 1 ml of sample under slight nitrogen pressure through a 3-mm column containing 1 g of alumina treated with silver nitrate. Then the column was heated to 75°C in a stream of nitrogen to remove the heptane, followed by nitrogen saturated with oct-1-ene (boiling point 45°C), which displaced the unsaturated compounds. Figure 47 shows the output as deduced from the gas density. The limit of detection was $10^{-3}\%$. Concentrations down to $10^{-6}\%$ were examined by trapping the first octene fraction, which was assayed quantitatively for hex-1-ene, hept-1-ene, and hept-2-ene by gas—liquid chromatography.

An ester mixture has similarly been separated on a charcoal column.

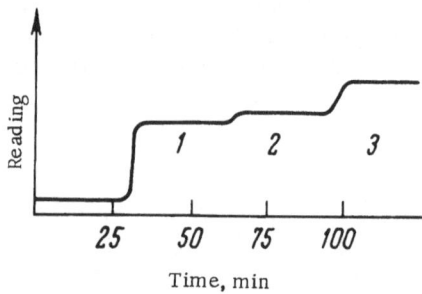

Fig. 47. Displacement of a concentrate of cis-heptene and trans-heptane by oct-1-ene in nitrogen (250 ml/min) on an alumina column treated with silver nitrate [107]: 1) trans-hept-3-ene, 2) cis-hept-3-ene, 3) oct-1-ene.

Time, min

Displacement enrichment has been used [108] for impurities in absolute ethanol with a 390×2 mm silica gel column, the principle being that alcohol is adsorbed more strongly than most other organic compounds, so all organic impurities apart from methanol should accumulate ahead of the ethanol. The absolute ethanol (0.8 ml) was injected at the top of a liquid column containing ASM silica gel (0.07-0.10 mm, 0.75 g). This was followed by water, and the first drop of eluate (6 μl) was analyzed by gas chromatography on a celite-545 column bearing 10% of grade 400 polyethylene glycol, length 2.7 m, at 50°C with high-purity nitrogen, the detector being of flame-ionization type. Figure 48 shows curves for the initial alcohol (scale 10^{-9} A) and the concentrate (scale 10^{-8} A, benzene at $3 \cdot 10^{-8}$ A). The limit of detection was 10^{-7}% and the coefficient of variation was 15%.

Simple and reliable apparatus is essential to successful liquid or gas chromatography. For instance, the effluent from a liquid column may be examined qualitatively and quantitatively drop by drop by gas chromatography [109, 110]. The capillary end of the liquid column allowed the liquid to fall drop by drop onto a heated grid. Solvent evaporation during drop growth was prevented by placing the capillary end in a cooled area. The vapor from the hot grid passed with a carrier gas to a gas-chromatography column, where it was separated into its components before passing to the detector. Continuous analysis is possible if the gas stage of analysis takes less time than the formation of the next drop.

Liquid chromatography has been used to enrich traces of sulfur compounds [111], fatty acids [112], etc.

Thin-layer chromatography (analytical or preparative) may be used to enrich compounds with very high boiling points [113, 114]; e.g., group separation may be performed by thin-layer chromatography, and then the individual fractions may be examined by gas chromatography [115]. Silica gels of KSK and ShSM types may be used; the first separates hydrocarbons well, while the second is better at separating aromatics and oxygen compounds. Plates 24×24 cm were used with 2-mm layers of the 0.05-0.16-mm fraction, with development by 40-60° petroleum ether. The two edges were left free for subsequent running of standards (naphthalene and n-nonan-5-one). The specimen was light oil from tunnel ovens (0.5 g), which was separated on the thin layer into four groups of

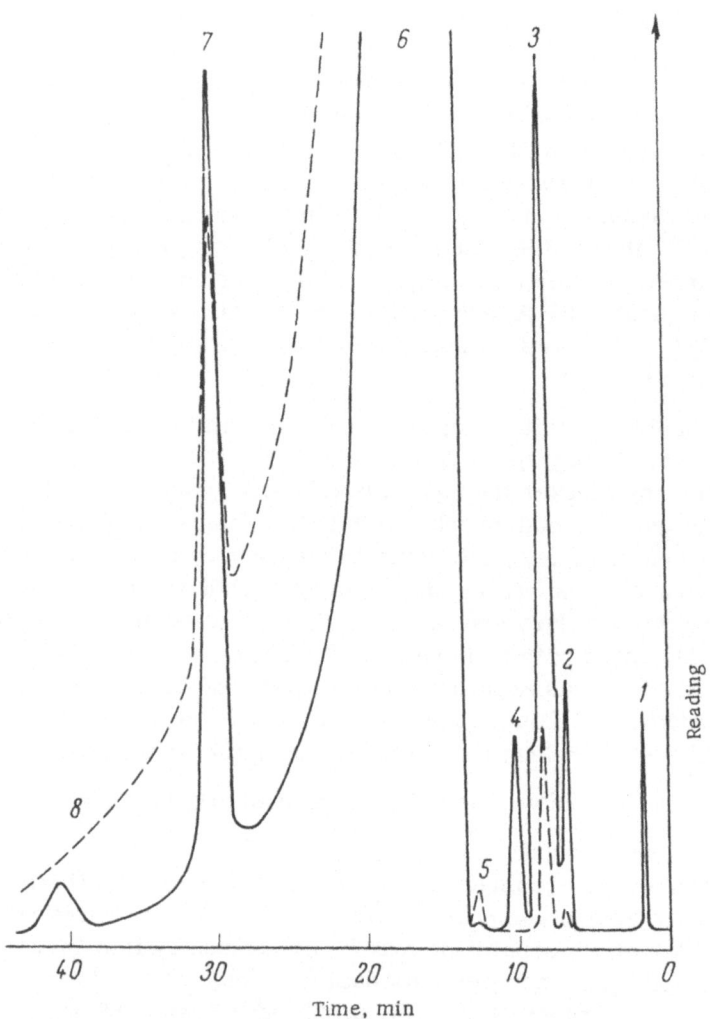

Fig. 48. Impurities assayed in absolute ethanol directly (broken line) and after enrichment by liquid chromatography (solid line): 1) diethyl ether, 2) ethyl acetate, 3) benzene, 4) unidentified, 5) methanol, 6) ethanol, 7) propanol, 8) unidentified.

compounds: alkanes, olefins, aromatics, and oxygen compounds. Diethyl ether and acetone were used to recover the fractions, which were then examined by gas chromatography.

Although this method was used to separate compounds present in comparable amounts, we can see no reason why it should not be applied to trace components, e.g., aromatic hydrocarbons in alkanes or traces of oxygen compounds in hydrocarbons.

Not much use has yet been made in gas chromatography of the considerable scope offered by liquid chromatography in impurity analysis and enrichment.

2. OTHER TECHNIQUES

Here we consider enrichment methods used in conjunction with analysis by gas chromatography. Of course, any selective method of separation may be the best in particular circumstances, e.g., distillation, zone melting, etc.

Absorption and condensation form the basis of the methods most frequently used for enrichment. A trap contains a liquid or is cooled to the temperature where the impurities condense, and through it is passed a measured volume of the sample (perhaps many liters), the main component passing freely through.

The early studies on preliminary impurity enrichment employed low-temperature condensation in empty U-tubes or ordinary sintered-glass filters; e.g., atmospheric contaminants were assessed [116] by condensation on filters cooled in liquid oxygen, with mass-spectrometric analysis and identification of the products. A limit of $10^{-8}\%$ was attained by using sample volumes up to 100 liters, and a further improvement by a factor 100 was possible if larger volumes were used. About 60 components were determined and identified in this way. A similar method was used [116a] to determine traces of water in hydrocarbon gases.

Selective sorption of the main component has been used [45] to determine trace impurities in CO_2, either by condensation at $-196°C$ or by absorption in a silica gel column. The hydrogen, oxygen, nitrogen, methane, and CO passed through to a column at 50°C containing 5A or 13X molecular sieve, with argon or helium as the carrier (the temperature could be reduced to $-80°C$ if Ar and O_2 had to be determined separately). The limits of detection were $0.1-2 \cdot 10^{-4}\%$ with the first method of removing the CO_2. The silica-gel method is more rapid, but the limits rise to $30-100 \cdot 10^{-4}\%$.

Enrichment is generally used in analysis for heavy impurities in air and other moist gases. A complication arises from the condensation of water in the trap at low temperatures, which markedly reduces the adsorption capacity, so a device for selective retention or conversion of the water is placed before the trap. There is difficulty in choosing the drying agent, since it should retain selectively only water vapor. Farrington et al. [117] examined various drying agents for analysis of various classes of compound. Table 12 gives the adsorption characteristics in relation to various compounds, which shows that KOH is the best, while ascarite passes nonpolar components almost completely but partly retains oxygen compounds [118]. West et al. [55] showed that magnesium perchlorate can be used, but polar compounds have to be recovered by heating the trap to 100°C. The trap temperature must be 150-180°C below the boiling point of any compound to be recovered, but

TABLE 12. Adsorption Activity of Various
Drying Agents*

Compound	4A molecular	Calcium sulfate	Magnesium perchlorate	Barium perchlorate	KOH
Isopropyl ether	25	100	100	75	100
Acetone	0	40	5	10	90
Isobutyraldehyde	10	50	75	25	100
Isopropanol	0	10	0	50	75
n-Butyraldehyde	0	50	75	25	100
Methyl ethyl ketone	5	20	5	25	100
Isopropyl acetate	5	0	5	25	100
Allyl ether	0	50	75	5	100
Isobutyl formate	0	<5	25	10	100
n-Propyl acetate	0	<5	—	<5	100

* The table states the proportion (%) passed by the reagent.

an excessively low temperature may produce an aerosol not retained by the trap. Hence liquid nitrogen, for example, is suitable
only for compounds that boil below 100°C.

Sometimes high-boiling compounds are taken up in a trap containing a suitable solvent not supported on a solid; e.g., high-boiling hydrocarbons in fuel gas were determined [119] with a cooled
trap containing n-octane, while Yavorovskaya [56] used ethanol or
amyl alcohol to trap chlorine compounds. Only a small fraction of
the solvent is taken for analysis in such cases [119a].

The above methods are applicable mainly to high-boiling compounds in gaseous materials. Sometimes one has to assay low-
boiling impurities in a relatively high-boiling base, for which purpose two methods have been described. The first employs prefer-
ential sorption of the principal component in a suitably filled trap
(see Section 1 of this chapter), while the second employs bubble
systems (various designs). A large sample is placed in the bubble
unit, and a gas stream is passed to remove the volatile impurities,
e.g., in the assay of oxygen in lubricating oils [1], where the vessel 12 mm in diameter had two sintered-glass discs, with injection
of the sample through a rubber seal, the helium carrier being
passed at 50 ml/min. The dissolved oxygen passed to a 600×0.6
cm column containing 5A molecular sieve. The entire system was
maintained at 75°C. The tube was shut off with three-way stopcocks after extraction. The method was made quantitative by calibrating the detector with pure oxygen.

This method causes the base to enter the chromatography column along with the oxygen, and the base usually poisons the column, which means that either regeneration or refilling will be
needed after a few runs. The same disadvantages occur in other
methods that use the same principle [120, 121].

A system has been proposed [3] for increasing the working
time of the forecolumn while using large volumes of liquid (10 ml
or more), which provides improved sensitivity.

If the impurities do not differ from the base in properties so
markedly as the above methods would require, the scope for enrichment by blowing a gas through the liquid is restricted largely
by the finite rate of extraction of the impurities from the sample.
Sometimes this difficulty is overcome by using intermediate en-

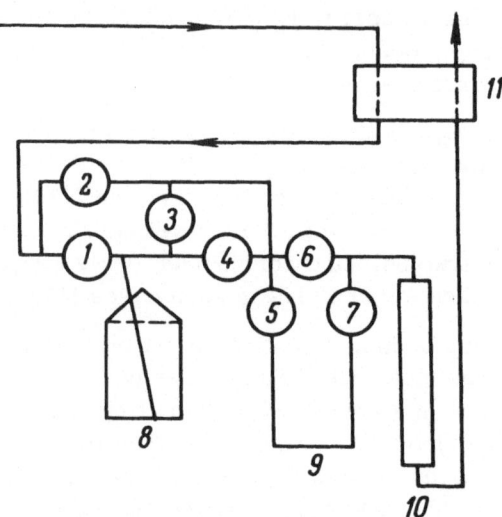

Fig. 49. Apparatus for assaying compounds with
intermediate enrichment [122]: 1-7) stopcocks,
8) sample vessel, 9) trap, 10) analytical column,
11) katharometer.

richment before the chromatography column, much as in the sorption method described above. Durrett.[122] used such a system to
assay methyl ethyl ketone and toluene in lubricating oil (Fig. 49).
The specimen (10 g) was placed in a steel cylinder at 220°C. The
30×0.6 cm trap was filled with 20-30 mesh diatomite at −50°C
(isopropanol + dry ice). The helium was passed through the sample and trap with stopcocks 1, 4, 5, and 7 open, while, 2, 3, and 6
were closed. The helium was passed through the oil at 220°C for
about 20 min, and then the cylinder and trap were isolated (stopcocks 2, 3, and 6 open, while 1, 4, 5, and 7 were closed). The cooling mixture was removed and the trap was heated to 125°C and kept
at that temperature for a few minutes, after which the desorbed
compounds were carried by the helium (40 ml/min) into the 300×
0.6 cm analytical column, which was filled with C-22 celite bearing Carbowax-22 (40:100) at −100°C. Water did not interfere with
the determination, because it was eluted later than the relevant
impurities. Internal standards were used for calibration by dis-
solving known quantities of the compounds in pure lubricating oil.

The limit of detection was $10^{-3}\%$ for the ketone when a katharom-eter was used, and $5 \cdot 10^{-2}\%$ for toluene.

Intermediate enrichment has also been used by Tolk et al. [122a].

Fractional extraction of volatile components from food prod-ucts has been performed [123] by a series of traps with succes-sively lower temperatures in a closed circulation system (Fig. 50) containing an inert gas. The volatile impurities were taken up in a packed cold trap until the vapor pressure there attained the vapor pressure over the sample. The enrichment is controlled by adjusting the various temperatures.

Alekseeva et al. [52] used intermediate concentration in assay-ing hydrocarbons in recovered solvents used in purification of mon-omers by extractive distillation. The hydrocarbons were blown out of 5-10 ml of sample by hydrogen flowing at 80 ml/min and were ac-cumulated in an 18×0.4 cm cooled U-tube containing a sorbent. The solvent vapor was removed by 1% alkali on diatomite, which does not react with the impurities.

Solvent extraction as a means of enrichment has not yet been widely used in gas-chromatographic impurity analysis, although it has proved valuable in inorganic chemistry. The method is based on using a solvent whose partition coefficient K for the impurities is much larger than that for the base.

If the solutions are very dilute (as is the case in impurity analysis), the concentration ratio is constant and independent of

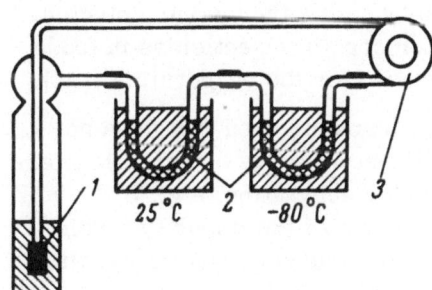

Fig. 50. Circulation system for fractional enrichment of volatile components [123]: 1) sample bottle, 2) traps at different tem-peratures, 3) circulation pump.

the amount of impurity, i.e.,

$$K = \frac{c_1}{c_2}.$$ (92)

Then the proportion of the impurity in the upper phase is

$$\Omega = \frac{Kr}{Kr+1},$$ (93)

where r is the volume ratio of the upper and lower phases. A countercurrent system gives the best scope for solvent extraction [124]. An example of a single-stage extraction is the determination of traces of cyclohexane in water [125], where 25 ml of water was treated with 1 ml of 2,2-dimethylbutane. The upper solvent layer was sampled with a syringe for gas chromatography. The method increases the cyclohexane concentration by a factor 25 and allows $7 \cdot 10^{-4}\%$ to be detected with a thermal-conductivity detector. An enrichment factor of 60-100 has been attained [131] for traces of petroleum in water by extraction into CCl_4. If necessary, several cycles of extraction with different solvents can be used, the volume being reduced at each stage [132].

Sometimes the limit of detection can be greatly improved by selective evaporation of the extracting solvent. This is especially effective if the impurity is unaffected by prolonged heating and if it does not form an azeotrope with the solvent, which can lead to low results. Evaporation has been used [133] to assay organophosphorus insecticides in water; to 1 liter of sample was added 10 ml of saturated NaCl solution, and the compounds were extracted into petroleum ether (100 and 50 ml), the extract being dried over sodium sulfate and evaporated to a final volume of 1 ml. The product was analyzed by gas chromatography with a thermionic detector. A similar method has been used for certain insecticides in foods [134]. A simple apparatus can be used for the evaporation [135].

Solvent extraction is probably desirable even if it does not produce enrichment if the extracting solvent has better characteristics as regards the chromatographic separation, e.g., if the retention is much greater or less than that of the impurity. This may allow the sample volume to be increased considerably, with the corresponding improvement in the limit of detection.

Solvent extraction is very promising when direct gas chromatography of the specimen is impossible, e.g., polymers. Lewis and Patton [126] examined extracts from plasticized press compositions to identify the components used in plasticization. It is possible to use solvent extraction to examine plastic weathering, to detect impurities in intermediate products, to assay plasticizers, etc. Bradley [136] assayed aldrin in fertilizers by extraction into hexane; herbicides have been extracted from commercial products [137]; and bromacyl ($5 \cdot 10^{-7}\%$) in soil has been assayed [138].

A pesticide (Melatiol) in wheat has [139] been assayed by solvent extraction in conjunction with additional enrichment by evaporation; 20 g of specimen was treated with 100 ml of acetone acidified with 0.5-1.0 ml of $2N$ HCl, the solution was filtered, and the analysis was performed by gas chromatography with an electron-capture detector. The limit of detection was 0.1-0.4 ng, or 0.05-0.2 ng when a detector selective for phosphorus was used. Substantial evaporation has been used [140] in the extraction of Eptane from soils with petroleum ether, which provided detection of 0.2 mg/kg on 200-300-g samples.

The scope for use of solvent extraction in gas chromatography is far from exhausted.

An important advantage of solvent extraction is that it can provide valuable information on the extracted compounds; the partition between the liquids often provides an efficient means of identification [127-130].

LITERATURE CITED

1. P. G. Elsey, Anal. Chem., 31:869 (1959).
2. J. Halasz and W. Schneider, Brennstoff-Chem., 41:225 (1960).
3. V. R. Alishoev, V. G. Berezkin, and V. P. Pakhomov, Zav. Lab., 32:1204 (1966).
4. A. A. Zhukhovitskii and N. M. Turkel'taub, Gas Chromatography [in Russian], Moscow, Gostoptekhizdat (1962).
5. S. D. Nogare and R. S. Juvet, Gas—Liquid Chromatography, Theory and Practice (1962).
6. K. I. Sukodynskii, N. A. Malafeev, and N. M. Zhavonkov, in: Gas Chromatography [in Russian], Moscow, GOSINTI (1962), p. 26.
7. K. V. Alekseeva, V. G. Berezkin, S. A. Volkov, and E. G. Rastyannikov, Production of Pure Substances by Preparative Gas Chromatography [in Russian], Moscow, TsNIITÉNeftekhim (1968).

8. M. Verzele, J. Chromat., 13:377 (1964).
9. J. Thompson, J. Chromat., 6:454 (1961).
10. F. L. Snelson, Chem. Ind., 575 (1964).
11. N. W. R. Daniels, Chem. Ind., 1078 (1963).
12. H. Schlenk and D. M. Sond, Anal. Chem., 34:1676 (1962).
13. R. Teranishi, J. W. Corse, J. C. Day, and W. G. Lennings, J. Chromat., 9:244 (1962).
14. R. K. Stevens and J. D. Mold, J. Chromat., 10:398 (1963).
15. R. Teranishi, R. A. Flath, T. R. Mon, and R. K. Stevens, J. Gas Chromat., 3:206 (1965).
16. P. A. J. Swoboda, Nature, 199:31 (1963).
17. R. Hardy and J. N. Klay, J. Chromat., 17:177 (1965).
18. V. G. Berezkin and L. S. Polak, Kinetika i Kataliz, 2:285 (1961).
19. H. W. Leggon, Anal. Chem., 33:1295 (1961).
20. J. M. Lesser, Anal. Chem., 31:484 (1959).
21. M. D. Howlett and D. Walti, Analyst, 91:291 (1966).
22. J. D. Boggus and N. G. Adams, Anal. Chem., 30:1471 (1958).
23. L. Mikkelsen, Second Pittsburgh Conference on Analytical Chemistry and Applied Spectroscopy, Pittsburgh, Pa., 1960.
24. B. K. Krylov and V. I. Kolmanovskii, Trudy po Khim. i Khim. Tekhnol. (Gor'kii), 4:742 (1961).
25. D. A. Vyaklirev, Z. S. Smolyan, P. B. Reshetnikova, I. D. Demina, and M. I. Vlasova, Trudy po Khim. i Khim. Tekhnol. (Gor'kii), 3:490 (1961).
26. J. Serpinet, J. Chim. Anal., 42:433 (1960).
27. E. Bayer, Chimia, 17:199 (1963).
28. E. Bayer, K. P. Hupe, and H. Mack, Anal. Chem., 35:492 (1963).
29. T. Johns, M. R. Burnell, and D. W. Carle, Gas Chromatography, Second International Symposium, 1959, edited by H. J. Noebels, R. F. Wall, and N. Brenner, New York, Academic Press (1961).
30. I. D. McCallum, Progress in Industrial Gas Chromatography, Vol. 1, New York, Plenum Press (1961).
31. D. Dinelly, S. Palerro, and M. Taramasso, J. Chromat., 7:477 (1962).
32. M. Taramasso, F. Sullasto, and A. Guerra, J. Chromat., 20:220 (1965).
33. V. G. Berezkin and E. G. Rastyannikov, Authors' certificate 194,408 (1966); Byull. Izobr., No. 8, 108 (1967).
34. V. G. Berezkin and E. G. Rastyannikov, Izv. AN SSSR, Ser. Khim., 240 (1968).
35. A. A. Zhukhovitskii, V. V. Naumova, M. S. Selenkina, N. M. Turkel'taub, V. P. Shvartsman, and A. F. Shlyakhov, Gas Chromatography, Proceedings of the Third All-Union Conference [in Russian], Izd. Dzerzh. Fil. OKBA (1966), p. 74.
36. A. A. Zhukhovitskii, N. M. Turkel'taub, L. A. Malyasova, A. F. Shlyakov, V. V. Naumova, and T. I. Pogrebnaya, Zav. Lab., 35:1162 (1963).
37. A. A. Zhukhovitskii, N. M. Turkel'taub, V. P. Shvartsman, and A. F. Shlyakhov, Dokl. AN SSSR, 150:654 (1964).
38. M. M. Dubinin and M. V. Khenova, Zh. Prikl. Khim., 19:1204 (1936).
39. M. M. Dubinin and S. Zh. Yavich, Zh. Prikl. Khim., 19:1021 (1936).

40. G. Schay, Theoretical Principles of Gas Chromatography [Russian translation], Moscow, IL (1963).

41. S. Classon, Adsorption Analysis of Mixtures [Russian translation], Moscow, Goskhimizdat (1949).

42. O. V. Al'tshuler, D. M. Vinogradova, S. Z. Roginskii, and Yu. I. Chirkov, Gas Chromatography, Proceedings of the Third All-Union Conference [in Russian], Izd. Dzerzh. Fil. OKBA (1966), p. 317.

43. O. V. Al'tshuler, D. M. Vinogradova, S. Z. Roginskii, and Yu. I. Chirkov, Dokl. AN SSSR, 152:892 (1963).

44. V. S. Mirzayanov, A. A. Zhukhovitskii, V. G. Berezkin, and N. M. Turkel'taub, Zav. Lab., 29:1166 (1963).

45. R. Aubeau and L. Chempeix, Ind. Atom., 4(11-12):78 (1960).

46. P. I. Markosov, V. N. Zaichenko, and Z. A. Lityaeva, Zav. Lab., 27:285 (1961).

47. V. G. Berezkin and V. S. Mirzayanov, Izv. AN SSSR, Ser. Khim., 1202 (1967).

48. V. G. Berezkin and V. S. Mirzayanov, Gas Chromatography, Proceedings of the Third All-Union Conference [in Russian], Izd. Dzerzh. Fil. OKBA (1966), p. 105.

49. N. M. Turkel'taub, L. N. Ryabchuk, S. N. Morozova, and A. A. Zhukhovitskii, Zh. Anal. Khim., 19:133 (1964).

50. V. S. Mirzayanov and V. G. Berezkin, Neftekhimiya, 4:641 (1964).

51. A. A. Datskevich and P. I. Avramenko, "Automation in technological processes," Trudy VNIIKANeftegaz, No. 2, Moscow, Nedra (1968), p. 196.

52. K. V. Alekseeva, V. K. Komarova, and N. P. Slepova, Gas Chromatography, No. 2, Moscow, NIITÉKhim (1964), p. 112.

52a. B. K. Krylov and V. I. Kalmanovskii, Zav. Lab., 35:152 (1969).

52b. A. A. Zhukhovitskii, N. M. Turkel'taub, and V. V. Naumova, Neftekhimiya, 6:324 (1966).

52c. R. A. Sanford and B. D. Ayers, U.S. patent, Class 55-67, No. 3,457,704, submitted October 10, 1966; registered July 29, 1969.

53. A. W. Mosen and G. Buzzelli, Anal. Chem., 32:141 (1960).

54. D. M. G. Lewrey and C. C. Cerato, Anal. Chem., 31:1011 (1959).

55. P. W. West, B. Sen, and N. A. Bibson, Anal. Chem., 30:1390 (1958).

56. S. F. Yavorovskaya, Trudy Kom. po Anal. Khim. AN SSSR, 13:269 (1963).

57. N. Brenner and L. S. Ettre, Anal. Chem., 31:1815 (1959).

58. F. R. Cropper and S. Kaminsky, Anal. Chem., 35:735 (1963).

58a. V. B. Silin, A. V. Markevich, and S. A. Dobizhin, Zh. Anal. Khim., 23:1506 (1968).

59. V. I. Kalmanovskii and A. A. Zhukhovitskii, J. Chromat., 18:243 (1965).

60. V. I. Kalmanovskii and A. A. Zhukhovitskii, Gas Chromatography, Proceedings of the Third All-Union Conference [in Russian], Izd. Dzerzh. Fil. OKBA (1966), p. 204.

61. N. L. Brenner, E. Cieplinski, L. S. Ettre, and V. J. Coates, J. Chromat., 3:230 (1960).

62. M. Beroza, J. Gas Chromat., 2:330 (1964).

63. J. Haslam, Analyst, 86:44 (1961).

64. M. D. Howlett and D. Walti, Analyst, 91:291 (1966).

65. D. A. Shearer, Analyst, 88:147 (1963).

66. F. T. Eggersten and F. M. Nelsen, Anal. Chem., 30:1040 (1958).
67. J. H. Williams, Anal. Chem., 37:1723 (1965).
68. A. A. Carlstrom, C. F. Spencer, and J. F. Johnson, Anal. Chem., 32:1056 (1960).
69. R. Aubeau, L. Champeix, and J. Reess, J. Chromat., 16:7 (1964).
70. A. Dravnieks and B. K. Krotaszynski, J. Gas Chromat., 4:367 (1966).
71. Al. Purer, J. Gas Chromat., 3:165 (1965).
72. C. L. Franst and E. R. Hermann, Amer. Industry Hyg. Ass. J., 27:68 (1966).
73. L. A. Ruslanova, V. S. Tatarinskii, G. I. Volkova, V. G. Berezkin, A. K. Zhomov, and N. I. Kholdyakov, Nauch. Rab. NII Okhrany Truda, 51:66 (1968).
74. J. Novak, V. Vasak, and J. Janák, Anal. Chem., 37:661 (1965).
75. A. A. Zhukhovitskii, N. M. Turkel'taub, and T. V. Georgievskaya, Dokl. AN SSSR, 92:987 (1953).
76. A. A. Zhukhovitskii and N. M. Turkel'taub, Zav. Lab., 22:1952 (1956).
77. N. M. Turkel'taub and A. A. Zhukhovitskii, Geologiya Nefti, No. 2, 54 (1957).
78. E. V. Vagin and N. M. Garbuzov, Gas Chromatography, Proceedings of the Second All-Union Conference [in Russian], Moscow, Nauka (1964), p. 367.
79. E. V. Vagin and S. S. Petukhov, Gas Chromatography, Proceedings of the Second All-Union Conference [in Russian], Moscow, Nauka (1964), p. 125.
80. S. S. Petukhov and E. V. Vagin, in: Gas Chromatography [in Russian], Moscow, GOSINTI (1960), p. 97.
81. R. Kaiser, Z. Anal. Chem., 236:168 (1968).
82. H. Pauschmann, Z. Anal. Chem., 236:159 (1968).
83. H. Oster, Siemens-Z., 42:703 (1968).
83a. R. Kaiser, Chromatographia, No. 5-6, 199 (1968); No. 10, 453 (1969).
84. A. A. Zhukhovitskii and N. M. Turkel'taub, Gas Chromatography, Proceedings of the First All-Union Conference [in Russian], Moscow, Izd. AN SSSR (1960), p. 107.
85. J. J. Kirkland, ACS Anal. Summer Symp., Houston, Tex., June 1960, p. 14.
86. B. S. Tul'chinskii, L. K. Dereza, and L. I. Tul'chinskaya, Authors' certificate 181,045 (1966).
87. V. R. Alishoev, V. G. Berezkin, and V. P. Pakhomov, Izv. AN SSSR, Ser. Khim., 686 (1967).
88. C. Henjes, Öl und Köhle, 14:1079 (1938).
89. N. Turner, National Petr. News, 35:234 (1943).
90. M. I. Yanovskii, Ph.D. Thesis, Institute of Physical Chemistry, Academy of Sciences of the USSR, Moscow (1947).
91. E. V. Vagin and A. A. Zhukhovitskii, Dokl. AN SSSR, 94:973 (1954).
92. M. I. Yanovskii, S. P. Oziranev, and Lu P'ei-chang, Zh. Prikl. Khim., 33:1084 (1960).
93. E. V. Vagin, Gas Chromatography, Proceedings of the First All-Union Conference [in Russian], Moscow, Izd. AN SSSR (1960), p. 118.
94. S. S. Petukhov and E. V. Vagin, Gas Chromatography, Proceedings of the First All-Union Conference [in Russian], Moscow, Izd. AN SSSR (1960), p. 292.
95. O. V. Al'tshuler, O. M. Vinogradova, V. R. Linde, S. Z. Roginskii, Yu. N. Chirkov, and M. I. Yanovskii, Gas Chromatography, Proceedings of the Second All-Union Conference [in Russian], Moscow, Nauka (1964), p. 150.

96. O. V. Al'tshuler, O. M. Vinogradova, S. Z. Roginskii, and M. I. Yanovskii,
     Dokl. AN SSSR, 140:307 (1961).
97. S. Z. Roginskii, O. V. Al'tshuler, O. M. Vinogradova, M. I. Yanovskii, and
     O. P. Krivoruchko, Izv. AN SSSR, Ser. Khim., 214 (1965).
97a. A. A. Zhukhovitskii, N. M. Turkel'taub, and V. V. Naumova, Neftekhimiya,
     6:324 (1966).
98. V. Yu. Zel'venskii and K. K. Sakodynskii, Gas Chromatography, No. 5, Moscow,
     NIITÉKhim (1967), p. 77.
99. E. Glückauf and G. P. Kitt, Vapour Phase Chromatography Symposium, Butter-
     worths, London (1957), p. 422.
100. O. V. Al'tshuler, O. M. Vinogradova, S. Z. Roginskii, and Yu. N. Chirkov, Dokl.
     AN SSSR, 152:892 (1963).
101. O. V. Al'tshuler, O. M. Vinogradova, S. Z. Roginskii, and Yu. N. Chirkov, Gas
     Chromatography, Proceedings of the Third All-Union Conference [in Russian],
     Izd. Dzerzh. Fil. OKBA (1966), p. 317.
102. V. G. Berezkin and E. G. Rastyannikov, Neftekhimiya, 6:487 (1966).
103. V. G. Berezkin and E. G. Rastyannikov, Izv. AN SSSR, Ser. Khim., 749 (1969).
104. S. Dal Nogare and L. W. Safranski, J. Chem. Ed., 35:14 (1958).
105. P. M. Baxter, P. C. Dandya, S. I. Kandel, A. Okany, and J. C. Walker, Nature,
     185:466 (1960).
106. J. W. White, Jr., M. K. Reader, and M. L. Riethof, J. Assoc. Offic. Agr.
     Chemists, 43:778 (1960).
107. E. G. Scott and C. S. G. Phillips, Nature, 199:66 (1963).
108. V. G. Berezkin, L. I. Kolomiets, and V. S. Tatarinskii, Zh. Anal. Khim.,
     24:1095 (1969).
109. V. R. Alishoev, V. G. Berezkin, and V. S. Tatarinskii, Zav. Lab., 34:148 (1968).
110. V. R. Alishoev, V. G. Berezkin, and V. S. Tatarinskii, Authors' certificate
     192,488 (1965); Byull. Izobr., No. 5, 125 (1967).
111. C. H. Amberg, J. Inst. Petrol., 45:1 (1959).
112. P. B. Hughes, Nature, 181:1281 (1958).
113. Thin-Layer Chromatography, edited by E. Stahl [Russian translation], Moscow,
     Mir (1965).
114. A. A. Akhrem and A. M. Kuznetsova, Thin-Layer Chromatography [in Russian],
     Moscow, Nauka (1964).
115. J. Klesment and A. Kasberg, Izv. AN SSSR, Ser. Khim., 17:258 (1968).
116. M. Schephert, S. M. Rock, R. Howard, and J. Stormes, Anal. Chem., 23:1431
     (1951).
116a. N. M. Turkel'taub, B. M. Luskina, and K. A. Palomirchuk, Zh. Anal. Khim.,
     22:1089 (1967).
117. P. S. Farrington, R. L. Pecsok, R. L. Meeker, and I. G. Olson, Anal. Chem.,
     31:1512 (1959).
118. H. G. McKee, G. W. Rhoades, and W. A. McMahon, 136 ACS Meeting, Division
     of Water, Sewage, and Sanitation, Atlantic City, N.J., Sept. 1959, p. 13.
119. L. M. Kontorovich and V. P. Bobrova, Trudy po Anal. Khim. AN SSSR,
     13:257 (1963).
119a. D. A. Levaggi and M. Feldstein, J. Air Pollut. Control Ass., 19:43 (1969).

120. J. A. Petrocelli and D. H. Lichtenfells, Anal. Chem., 31:2017 (1959).

121. A. A. Kilner and J. A. Ratchiff, Anal. Chem., 36:1615 (1964).

122. L. R. Durrett, Anal. Chem., 31:1824 (1953).

122a. A. Tolk, W. A. Lingerak, A. Kont, and D. Borger, Anal. Chim. Acta, 45:137 (1969).

123. D. A. M. Mackay, ACS Anal. Summer Symp., Houston, Tex., June (1960), p. 14.

124. C. Craig, Trace-Element Analysis [Russian translation], Moscow, IL (1961), p. 106.

125. S. D. Nogare and L. W. Safranski, J. Chem. Ed., 35:14 (1958).

126. J. S. Lewis and H. W. Patton, Gas Chromat., N. Y., (1958), p. 145.

127. M. Beroza and M. C. Bowman, Anal. Chem., 37:291 (1965).

128. M. Beroza and M. C. Bowman, J. Assoc. Offic. Agr. Chemists, 48:358 (1965).

129. M. Beroza and M. C. Bowman, J. Assoc. Offic. Agr. Chemists, 48:943 (1965).

130. M. Beroza and M. C. Bowman, J. Assoc. Offic. Agr. Chemists, 48:493 (1965).

131. R. Jeltes, Water Res., 3(12):931 (1969).

132. Sueo Nishi and Yosijuki Horimoto, Tokyo Kogyo Shikensho Hokoku (Rep. Govt. Chem. Ind. Res. Inst., Tokyo, 64(2):493 (1969).

133. Gunter Zweig and James M. Devine, Residue Revs. No. 26, Berlin-Heidelberg-New York (1969), pp. 17-36.

134. J. Bäumler and S. Rippstein, Mitt. Geb. Lebenmitteluntersuch. und Hyg., 60(3): 171 (1969).

135. L. Kaziak and E. E. Hrast, J. Gas Chromat., 6(11):567 (1968).

136. J. K. Bradley, Chem. Ind. (London), 1976 (1961).

137. H. G. Higson and D. Butler, Analyst, 85:657 (1960.

138. Arthur Bevenue and James N. Ogata, J. Chromat., 46(1):110 (1970).

139. A. M. Kadoum, J. Agr. Food Chem., 17(6):1178 (1969).

140. R. E. Hughes and V. H. Freed, J. Agr. Food Chem., 9:381 (1961).

Chapter VIII

Some Aspects of Quantitative Analysis

An important point in impurity analysis is quantitative inter-pretation to give the amounts of the components in the mixture.

The accuracy is governed by various factors, including the method of analysis, the detector characteristics, the calibration and calculation methods, and the nature of the components. The detector and recorder are essentially to provide quantitative out-put, so we must consider briefly the relation between the concen-tration (amount) of a component and the parameters of the re-corded peak.

Novak [1] has shown that a concentration or flux detector gives a linear relation between the amount n_i of a component in a band and the peak area S_i:

$$n_i = c \frac{S_i}{\alpha_i - \alpha_0} = K_n S_i, \qquad (94)$$

where α_i and α_0 are the analytical parameter* values for the com-ponent and the carrier gas, while c is a factor dependent on the de-tector characteristics and recording conditions, and K_n is a coef-ficient of proportionality. In general, to deduce n_i we need to know S_i (or some other peak parameter) and K_n or some corresponding quantity.

Plate theory [3, 4] shows that the component concentration c is related to the gas volume V passed through the column by

---

*The analytical parameter is the physical or physicochemical property used in the analysis, e.g., thermal conductivity or heat of combustion [2].

$$c = \frac{N^{1/2}}{V_N} \cdot \frac{q}{(2\pi)^{1/2}} \exp \left\{ -\frac{N}{2} \left(1 - \frac{V}{V_N} \right)^2 \right\}, \tag{95}$$

where q = cw is the volume of the substance in the sample, whose total volume is w, N is the number of theoretical plates for the given separation conditions (temperature, gas speed, q, etc.), and V_N is the retention volume.

If the detector and recorder do not distort the curve, the latter faithfully reproduces the concentration curve, and the equation for a band is

$$h = \frac{q}{V_N} \cdot \frac{N^{1/2}}{\sqrt{2\pi}} \exp \left\{ -\frac{N}{2} \left(1 - \frac{x}{x_{max}} \right)^2 \right\} = h_{max} \exp \left\{ -\frac{N}{2} \left(1 - \frac{x}{x_{max}} \right)^2 \right\}, \tag{96}$$

where h is peak height, x is the chromatogram coordinate (reckoned from the point of injection), and x_{max} is x for $h = h_{max}$. It is best to choose as the quantity characterizing the amount one that satisfies the following requirements: (1) linear dependence on c, (2) independence of other components, (3) simplicity of measurement, (4) good reproducibility.

We now consider (96) on the assumption that a gaussian curve actually applies:

1. The peak area S is most often determined as the product of h_{max} and the width at half height $\mu_\theta = \mu_{1/2}$ ($\theta = c/c_{max} = \frac{1}{2}$) [5]

$$S = 0.941 h_{max} \mu_\theta. \tag{97}$$

The following is [6] a more general expression, which allows one to calculate the areas of partially separated peaks:

$$S = K_\theta h_{max} \mu_\theta. \tag{98}$$

The K_θ corresponding to θ of 0.5, 0.75, and 0.9 are 0.941, 1.66, and 2.73.

2. S may be determined from the product of h_{max} and retention time [1]:

$$S = \frac{2.21}{\sqrt{N}} h_{max} x_{max}. \tag{99}$$

3. The area is proportional to h_{max}:

$$S = \frac{2.21\, x_{max}}{\sqrt{N}}\, h_{max} = K_h h_{max}. \tag{100}$$

Then the amount is directly proportional to $h_{max}\mu_\theta$, $h_{max} x_{max}$, or h_{max}. In the first method, the coefficient of proportionality between the amount and the measured quantity is determined mainly by the detector characteristics and operating conditions, whereas in the second and third it is dependent also on the conditions in the chromatographic column. In particular, S is dependent on N in the second method, and N generally varies between substances [7]. In the third method, the coefficient of proportionality is dependent on the retention and the separation performance.

There are deviations from the linear relationships of (97)–(100) when the samples are large. For instance [8], the h_{max} for Xe at first increase in proportion to q, but deviations occur for samples greater than 2 ml, although the peaks remain symmetrical and the retention time is unaltered. The deviations are due to change in column performance and are to be expected whenever we have a value greater than 0.4 for the ratio of q to peak width (in volume units) for q very small [9]. The width increases with V_N, so the maximum q deduced from this inequality is not constant, and nonlinear effects can appear even for small q if V_N is small.

It is important to test for any dependence of the measured quantities on the working parameters, especially T and the gas speed. A study has been made [10] of the effects of T on h_{max} and S. Table 13 gives h_{max} and S as influenced by T in gas–liquid chromatography; S varies only slightly with T, while h_{max} increases by 1.5–4.1%/°C, so exact analyses via h_{max} require control of T closer than that needed when S is used.

Also, h_{max} is dependent on the sample evaporation rate and hence on the temperature of the injection device [11]. Figure 51 shows this. Stable results require the injector to be at a temperature somewhat above the boiling points of the compounds (dichloroethane 40°C, acetone 56°C). Excessive temperatures present the hazard of decomposition in the injector [12].

The gas speed w affects h_{max} and S; $S \propto 1/w$ [4], so that $dS/dw \propto 1/w^2$, and S is less dependent on w at high w.

TABLE 13. Effects of Temperature on Peak
Height and Area* [13]

Compound	h_{max}, mV/mg			S, ml-mV/mg		
	80° C	105° C	Change, %/°C	80° C	105° C	Change, %/°C
n-Pentane	4.14	5.74	1.5	134	145	0.3
n-Hexane	2.24	3.74	2.7	132	150	0.5
Cyclohexane	1.21	2.20	3.3	118	133	0.5
n-Heptane	1.02	2.06	4.1	122	141	0.6
Methyl-cyclohexane	0.70	1.39	3.9	107	124	0.6

* The measurements were made with a column 305×0.635 cm containing 40% squalane on a solid carrier, gas flow rate 40 ml/min.

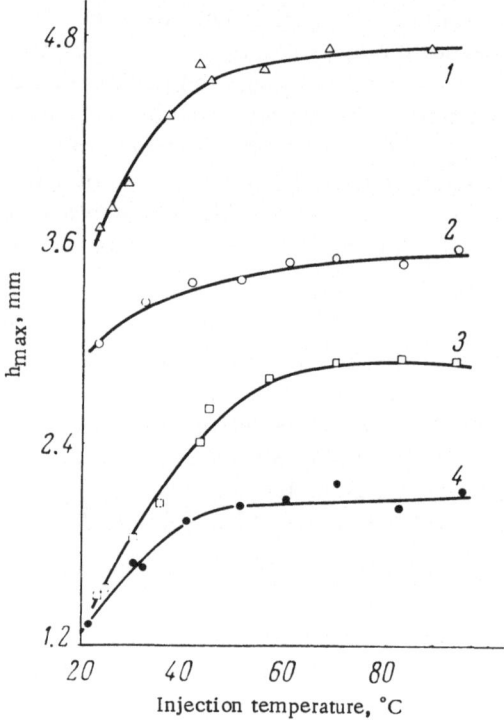

Fig. 51. Effects of injection temperature on peak
height [11]. Dichloroethane: 1) 3-μl sample,
2) 2-μl sample; acetone: 3) 3 μl, 4) 2 μl.

On the other hand, h_{max} is only slightly dependent on w for w small, and it starts to decrease linearly as w increases [4, 13].

It is clear that S is less dependent on the working conditions than is h_{max}, which is why S is most often used. The following operations are involved in deducing S from the recording: measurement of h_{max} (from baseline), determination of the intermediate height to measure the width at $h_\theta = \theta h_{max}$, measurement of the width μ_θ at h_θ, and calculation of S.

The errors of experiment in determining S in this way have been considered [14]. Each of the four operations introduces some error, and the total error arises from: (1) the error Δm in measuring the lengths (height and width), (2) the error Δb in locating the baseline, (3) the error Δh in the position of the point where the width is measured:

$$\frac{\Delta S}{S} = \frac{\Delta m_h}{h} + \frac{\Delta b}{h} + \frac{2\Delta h / \left(\frac{\partial h}{\partial x}\right)_{x=\theta h_{max}}}{\mu} + \frac{\Delta m_\mu}{\mu} = \frac{\Delta m_h + \Delta b}{h} +$$

$$+ \frac{2\Delta h / \left(\frac{\partial h}{\partial x}\right)_{x=\theta h_{max}} + \Delta m_\mu}{\mu}. \tag{101}$$

Some simplifying assumptions were made [14] in discussing how the relative error in S is affected by the retention time (Fig. 52). Most of the error arises from the width measurement for

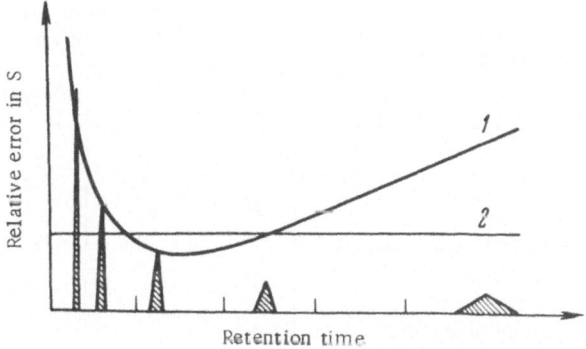

Fig. 52. Relative error in S as a function of retention time for peaks of constant S (five homologs) [14]. Method: 1) isothermal, 2) temperature program. The peaks are shown only for the first.

small times, while most of the error comes from uncertainty in h_{max} for large times.

The width measurement may be made more accurate by using a magnifier with an engraved scale (0.1-mm divisions [15]). Allowance must be made for the width of the recorded line. The width at a particular height ($\theta = 0.5$, $1/e$, etc.) may be measured with a special set of engraved curves [16]. If S is to be determined on narrow peaks, it may be best to measure the width for $\theta < 0.5$, while other peaks may be measured at $\theta = 0.5$ or even $\theta > 0.5$. There is a best ratio of height to width (the one giving the least error in S); this optimal ratio for $h_{max}/\mu_{\theta=1/3}$ varies [16] from 5 to 10.

The following methods are used to determine S for peaks of any shape, including unsymmetrical ones: (1) planimetry, (2) deduction of S by weighing of the cut-out curve, (3) integration in various ways.

Janák [17] used 100 analyses in a statistical evaluation of these methods. Table 14 gives the coefficients of variation for various S. The best methods are use of integrators and h_{max}.

TABLE 14. Statistical Evaluation of the Errors of Measurement of S by Various Methods [17]

Method	Mean coefficient of variation (%) for S (mm²) of			
	0-50	51-100	101-300	301-1200
Planimetry	40	9.1	5.0	5.8
Planimetry repeated five times	23	5.1	2.3	2.6
Weighing of cut-out peak	17	10	2.8	2.5
Area measurement on fitted triangle	17	9.5	3.4	0.9
Deduction as the product of the height and the width at half height	19	12	4.3	2.4
Mechanical integrator (one pulse = 5 mm²)	36	11	2.4	2.0
Electromechanical integrator (one pulse = 2 mm²)	10	3.0	2.7	1.3
Analog electronic integrator (eight pulses = 1 mm²)	2.2	1.7	0.9	1.3
Deduction of S from h_{max}	4.0	3.5	1.9	1.9

TABLE 15. Effects of Temperature on Relative
Peak Heights and Areas with
Cyclohexane as Standard

Compound	Relative peak height			Relative area		
	80° C	105° C	Change %/°C	80° C	105° C	Change %/°C
n-Pentane	3.41	2.6	—0.95	1.13	1.09	—0.14
n-Hexane	1.85	1.7	—0.32	1.12	1.13	+0.035
Cyclohexane	0.843	0.91	+0.32	1.03	1.06	+0.078
n-Heptane	0.578	0.632	+0.37	0.908	0.943	+0.154
Methylcyclo-hexane	1.0	1.0	0.0	1.0	1.0	0.0

It is stated [18] that the best accuracy is obtained in manual
processing if S is determined as the product of h_{max} and the width
at half height; triangulation and ordinary planimetry give less satisfactory results.

Tests were done in nine different laboratories on a fixed mixture with an internal standard [19], and it was found that the height
was the best parameter for quantitative evaluation [20].

Deans [21] found that the ratio to an internal standard used
with h_{max} gave a coefficient of variation less than that from use
of integrators by a factor 3-10. This at first sight conflicts with
the above results on the dependence of h_{max} and S on the working
conditions, but it should be remembered that: (1) modern instruments provide highly stable conditions, (2) the effects on the individual quantities tend to cancel out as regards the relative quantities used in the internal-standard method. For instance, Table 15
and Fig. 53 give our calculated values derived from [4, 10] for the
relative heights and areas as functions of column temperature and
gas speed. Table 16 compares the use of heights and areas in quantitative work [19]. Usually, especially in routine work, it is best to
use heights.

The following is the relation of amount of material to the peak
parameters and the detector characteristics:

$$n_i = \frac{0.941c}{(\alpha_i - \alpha_0)} h_{max} \mu_{1/2} = \frac{2.21c}{(\alpha_i - \alpha_0)\sqrt{N}} h_{max} x_{max} \ \frac{cK_n}{\alpha_i - \alpha_0} h_{max}. \quad (102)$$

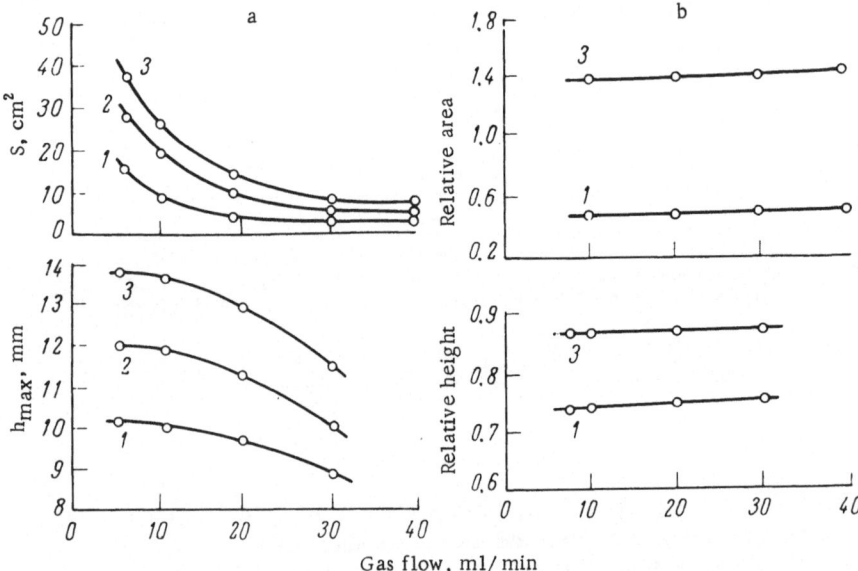

Fig. 53. Effects of gas flow on h_{max} and S: a) actual values, b) relative values; 1) ethyl formate, 2) ethyl acetate, 3) ethyl propionate.

TABLE 16. Comparison of Processing Methods

Calculation method	Advantages	Disadvantages
1. From areas	1. Linear dependence on concentration (amount) over a wide range 2. Scope for using published or calculated correction factors for detector sensitivity in calculating amounts	1. Reproducibility usually worse than for heights
2. From heights	1. Good reproducibility 2. Simplicity of measurement 3. Only a small effect on the heights from incomplete separation of peaks	1. Overloading results in nonlinear calibration 2. No possibility of using published sensitivity correction factors

It is therefore inadequate to determine only S or other peak parameters in order to find n_i; one needs to know also the coefficient of proportionality, which is determined by the type of detector, working conditions, nature of the material, etc. A distinctive feature of impurity analysis is that normalization cannot be used in quantitative calculations because the content of the main component cannot be determined with sufficient accuracy.

In this case, quantitative interpretation is provided via absolute calibration and use of an internal standard. These methods have the following advantages: (1) it is possible to measure only the relevant components, (2) it is possible to use mixtures containing components that are not recorded, (3) detector nonlinearity is balanced out, (4) the error of measurement is related only to the component being measured.

Absolute Calibration. Here one measures for each relevant component the area (or other parameter) as a function of the absolute amount, i.e., one of the following relationships is calibrated:

$$n_i = K_S S = K_h h = K_{h\mu} h_{max} \mu_\theta. \qquad (103)$$

This method is widely used [22-27], and Fig. 54 shows [22] the calibration curve for tetrafluoroethylene, whose concentration X

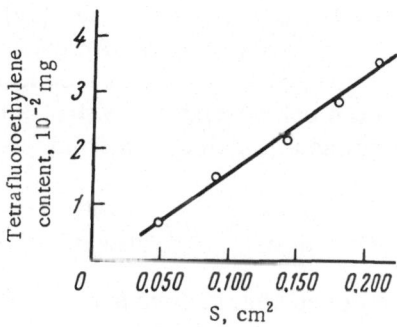

Fig. 54. Calibration curve for determining tetrafluoroethylene [22].

(mg/liter) is defined by

$$X = \frac{g}{V_0} ,$$ (104)

where g is the amount deduced from the calibration curve and V_0 is the volume of air taken for analysis, as reduced to NTP.

A modified form of the method uses a sample of constant volume provided by a dispensing system, in which case the calibration can be performed directly in % p:

$$p = K_{Sp}S = K_{hp}h = K_{h\mu p}h_{max}\mu_\theta.$$ (105)

This approach is simpler, and it gives good results for gases. Sometimes the difficulties found for liquids arise from dispensing error, which can be large for small samples (this problem is still not entirely solved). A density correction is needed if the results are to be presented in wt.%. Also, the absolute calibration is rather laborious, and the result is sensitive to the working conditions. However, nonlinearity is allowed for.

One can estimate the necessary constancy in T and gas flow rate from published evidence on the effects of these on h_{max}; constancy to 0.25°C and 1 ml/min reduce the relative error in h_{max} from these causes to 1% [28]. An absolute calibration should be checked periodically, with a frequency to be determined by experience. It is usually sufficient to check only certain points on the curve during repeat calibrations.

Internal-Standard Method. Ray [19] was the first to use this method in gas chromatography. An unknown mixture is treated to contain a standard substance at a concentration R, as reckoned relative to the entire mixture, which is taken as 100%. The content of a component is calculated from

$$p = \frac{f_i S_i R}{f_{st} S_{st}} \quad \text{or} \quad p = \frac{f_{hi} h_i R}{f_{hst} h_{st}},$$ (106)

where f_i and f_{st} are correction factors determined by the detector response to component i and the standard. If R is constant, one gets calibration curves whose coordinates are p and the ratio of peak height for i to peak height for the standard.

Relative calibration methods provide substantially increased accuracy because the working conditions have less effect on the results, since the standard and impurities are usually affected to identical extents, though there are exceptions here.

From (102) and (106) we get

$$\frac{p}{R} = \frac{n_i}{n_{st}} = \frac{f_i S_i}{f_{st} S_{st}} = \frac{f_i}{f_{st}} \cdot \frac{\sqrt{N_{st}}}{\sqrt{N_i}} \cdot \frac{V_i}{V_{st}} \cdot \frac{h_i}{h_{st}} = \frac{f_{hi}}{f_{hst}} \cdot \frac{h_i}{h_{st}},$$ (107)

where

$$f_i = \frac{1}{(\alpha_i - \alpha_0)}, \quad f_{st} = \frac{1}{(\alpha_{st} - \alpha_0)}.$$

These ratios are much less dependent on the working conditions than are S and h_{max}. Also, it is no longer necessary to have a strictly constant sample volume, except when calibration with a constant sample size is used in order to correct for detector nonlinearity. The method can be used even when not all the compounds are recorded. The standard must be compatible with the sample (must dissolve in it and not react with the components), and preferably it should be similar to the components in retention and in content in the mixture.

Internal standards are widely used in impurity analysis [4, 13, 28-31]. The method automatically corrects for possible losses of components in the preparative stage [29], e.g., when solvent extraction, distillation, adsorption, etc., are employed.

It has often been pointed out [4, 21, 32] that quantitative analysis by the internal-standard method is possible if the standard is not already present. This requirement restricts considerably the region of application, since one often has to examine complicated mixtures that cause difficulty in the choice of a standard.

In one form of the method [33], the internal standard is a substance present in the mixture, which has been used in impurity analysis [34].

Let $p_{st,1}$ be the concentration of the standard in the mixture and R be the concentration added (initial mixture taken as 100%), while p_i is the concentration of component i and $p_{st,1} + R = p_{st,2}$.

Quantitative analysis requires two chromatographic determinations: on the initial mixture and on the one with added standard. Then

$$\frac{p_{st,1}}{p_i} = \frac{f_{st}S_{st,1}}{f_i S_{i,1}} = K_{i,1},$$
(108)

$$\frac{p_{st,2}}{p_i} = \frac{p_{st,1} + R}{p_i} = \frac{f_{st}S_{st,2}}{f_i S_{i,2}} = K_{i,2}.$$
(109)

Then

$$p_i = \frac{R}{K_{i,2} - K_{i,1}}.$$
(110)

If the standard is not initially present, $K_{i,1} = 0$, and (110) becomes (106).

If a component is used as the internal standard, we have

$$p_{st} = \frac{R}{\dfrac{K_{i,2}}{K_{i,1}} - 1}.$$
(111)

A formula similar to (111) has been used elsewhere [30, 31]. In this particular case it is not necessary to know the correction coefficients.

The above equations are correct if the detector is linear. The contents of the other components are calculated via (112) and (113) via a chromatogram for the initial mixture:

$$p_i = p_{st} \frac{K_i S_{i,1}}{K_{st} S_{st,1}}$$
(112)

or a mixture containing the standard

$$p_i = (p_{st} + R) \frac{K_i S_{i,2}}{K_{st} S_{st,2}}.$$
(113)

Tests [34] show that this form of the method has advantages: the error is no greater than that in the widely used normalization method [4].

Sometimes, if a linear detector is used, one can use a hybrid method of quantitative analysis, where the content of an impurity (or perhaps of several) is deduced from an absolute calibration curve (provided this compound gives a well-resolved peak), while the other components are assayed by a method analogous to the internal-standard one via (112).

Standard-Mixture Method. This is best used with samples having a standard volume, and it requires a detector linear for all components. The standard mixture is run periodically during the analyses, this mixture having component concentrations close to those of the actual samples. (Analysis of similar mixtures is a typical problem in routine industrial laboratories.)

The component contents are then calculated from

$$p_i = \frac{h_i}{h_{i,\,st}}\, p_{i,\,st},$$ (114)

where h_i is the peak height for component i in the sample, $h_{i,st}$ is the same for the standard mixture, and $p_{i,st}$ is the content of component i in the standard mixture. If the mixture composition in some process has a specified composition, it is best to use a mixture having that composition as standard, and then the results can be presented as the relative quantities p:

$$\bar{p} = \frac{p_i}{p_{st}} = \frac{h_i}{h_{st}} = \frac{S_i}{S_{st}}.$$ (115)

Many of the above methods involve the assumption (1) that the components are not adsorbed or altered during the analysis, (2) that the detector response is linear over a wide concentration range and is unaffected by other components, and (3) that published correction factors are reliable.

Unfortunately, many of these conditions are not met in practice. In Chapter I we considered in detail the role of impurity adsorption on the solid carrier, and we showed that gross errors can arise. There are also papers that show that the detector response is affected by other components, in particular water [21, 35]. For instance, a flame-ionization detector was used to detect pentane in dry diglym by the internal-standard method (benzene content $1.2 \cdot 10^{-2}\%$), which gave a ratio of the pentane and benzene areas of 0.505, while a 50:50 mixture with water gave only 0.281 for this ratio [21]. Deans [21] found that the relative correction factors for hydrocarbons varied by 20-30% when use was made of chromatographs made by various firms (all had flame-ionization detectors). The variations were less for katharometers, but were still 3-6% of the published values [21]. None of the detectors examined in [21] was linear over a wide concentration range, so exact quantitative results can be obtained only if the instrument is calibrated with a mixture of known composition that contains all the compounds present in the actual samples. The calibration has to be checked periodically.

Statistical methods need to be more widely used [36, 37] to evaluate the errors. See [38, 39] for details of the techniques to be used.

LITERATURE CITED

1. J. Novak, Chem. Listy, 59:1021 (1965).
2. F. Gǔta, Přednašky o fysikalnich a specialich metodach analytickych, Prague, SNTL (1961).
3. A. J. P. Martin and R. L. M. Synge, Biochem. J., 35:1358 (1941).
4. S. D. Nogare and R. S. Juvet, Gas–Liquid Chromatography, Theory and Practice (1962).
5. E. Cremer and R. Müller, Z. Elektrochem., 55:217 (1951).
6. A. A. Zhukhovitskii, B. A. Kazanskii, O. D. Sterligov, and N. M. Turkel'taub, Dokl. AN SSSR, 123:1037 (1958).
7. J. H. Purnell, Nature, 184:2009 (1959).
8. B. G. Anstov, V. G. Keilal, A. V. Kiselev, and K. D. Shcherbakova, Gas Chromatography, No. 6, Moscow, NIITÉKhim (1967), p. 61.
9. V. I. Kalmanovskii and A. A. Zhukhovitskii, Gas Chromatography, Proceedings of the Third All-Union Conference [in Russian], Izd. Dzerzh. Fil. OKBA (1966), p. 93.
10. M. Dimbat, P. E. Porter, and F. H. Stross, Anal. Chem., 28:290 (1956).
11. F. H. Pollard and C. T. Hardy, Chem. Ind., 1145 (1955).
12. E. A. Day and P. H. Miller, Anal. Chem., 34:869 (1962).
13. E. Leibnitz and H. G. Struppe, Handbuch der Gas-Chromatographie, Leipzig, Akademische Verlagsgesellschaft (1966).
14. W. E. Harris and H. W. Habgood, Gas Chromatography with Temperature Programming [Russian translation], Moscow, Mir (1968).
15. V. G. Baranova, A. G. Pankov, and Ya. I. Tur'yan, Principles of Physicochemical Methods of Analysis and Monitoring in Isoprene Production [in Russian], Moscow, NIITÉKhim (1965).

16. D. L. Ball, cited in [9].
17. J. Janák, J. Chromat., 3:308 (1960).
18. R. P. W. Scott and D. W. Grant, Analyst, 89:179 (1964).
19. N. H. Ray, J. Appl. Chem., 4:21 (1954).
20. J. Chromat., 12:293 (1963).
21. D. R. Deans, Chromatographia, No. 5-6, 187 (1968).
22. N. Brenner and L. S. Ettre, Anal. Chem., 31:1815 (1959).
23. B. A. Rose, Analyst, 84:544 (1959).
24. H. Pitsch, Erdöl und Kohle, 11:157 (1959).
25. A. V. Alekseeva, V. P. Bobrova, and A. I. Fomina, Neftekhimiya, 5:449 (1965).
26. A. I. Dolgina, A. V. Alekseeva, and A. N. Meshcheryakova, Gas Chromatography, No. 4, Moscow, NIITÉKhim (1966), p. 121.
27. W. D. Ross and R. E. Sievers, Sixth International Symposium on Gas Chromatography and Associated Techniques, Rome (1966), Preprints.
28. A. A. Zhukhovitskii and N. M. Turkel'taub, Gas Chromatography [in Russian], Moscow, Gostoptekhizdat (1962).
29. H. P. Burchfield and Eleanor E. Storrs, Biochemical Applications of Gas Chromatography, New York, Academic Press (1962).
30. R. Kaiser, Gas Chromatography, Leipzig, AVG (1966).
31. G. Schay, Theoretical Principles of Gas Chromatography [Russian translation], Moscow, IL (1963).
32. A. J. M. Keulemans, Gas Chromatography, 2nd edition, edited by C. G. Verver, New York, Reinhold (1959).
33. V. G. Berezkin, I. A. Musaev, V. S. Tatarinskii, and P. I. Sanin, Gas Chromatography, No. 2, Moscow, NIITÉKhim (1965), p. 25.
34. I. P. Ogloblina, V. G. Makarenko, and V. A. Yarova, Gas Chromatography, No. 8, Moscow, NIITÉKhim (1968), p. 62.
35. I. S. Foster and J. W. Marein, Analyst, 90:118 (1965).
36. A. N. Zaidel', Elementary Estimates of Errors of Measurement [in Russian], Leningrad, Nauka (1967).
37. K. Doerfeld, Statistics in Analytical Chemistry [Russian translation], Moscow, Mir (1969).
38. R. Kaiser, Chromatographie in der Gasphase, Vol. IV, Mannheim (1965).
39. D. E. Dishina, Progress in Gas Chromatography [in Russian], No. 1. Mendeleev All-Union Chemical Society, Kazan' (1968), p. 91.

Chapter IX

Basic Methods of Preparing
Standard Mixtures

Standard mixtures of known composition are used in calibrating chromatographic devices and for periodically checking methods.

We have seen in Chapter III that the signal from a differential detector is very much dependent on the nature of the compound, so exact results require measurement of the detector response for each compound. The purpose of calibration is to provide a quantitative relation between the signal and the amount of a component, which may be produced with a standard mixture, which must be prepared with an accuracy in excess of that of the instrument. Preparation of standard mixtures containing trace components is therefore an important problem, which is made more difficult by the scope for entry of contaminants during preparation and use, as well as by adsorption, which results in change in the mixture composition.

Static methods of mixture preparation are most convenient for (1) gaseous or volatile substances handled in vapor form and (2) liquids prepared in solvents.

The simplest method is to use a gas-handling system of known volume [1, 2], e.g., a glass gas pipette (Fig. 55). A measured quantity of acetaldehyde is sealed into a tube and is placed in the pipette, which is sealed off, evacuated, and then filled with a previously prepared mixture of ethylene and oxygen. The tube is then broken and the mixture is stirred with a PTFE plate.

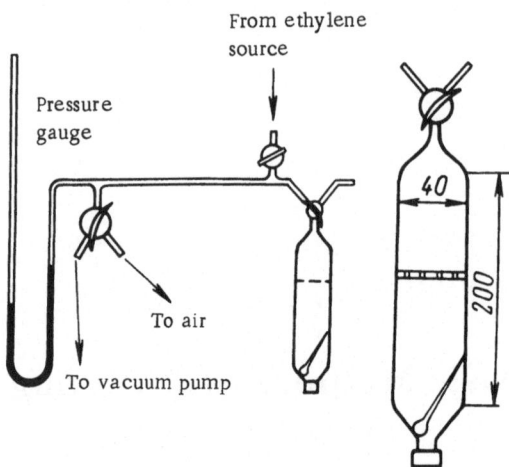

Fig. 55. Preparation of a calibration mixture in a
glass pipette [2].

A sample prepared in this way can be used directly in calibration or can be further diluted; in the latter case, the system is
evacuated to a set pressure and then refilled with the main component. The final concentration is calculated. In this way one can
produce concentrations down to $10^{-3}\%$ by volume [3]. The level
$0.5-5 \cdot 10^{-4}\%$ can be reached [4] by introducing a known amount of
material into a glass system and then flushing the latter with an
appropriate volume of gas into a previously evacuated vessel (complete evaporation is produced by heating before the flushing, if necessary). Plastic vessels may be used to reduce adsorption [4].

These methods can be used only when very small amounts of
mixture are needed for calibration. Larger volumes are prepared
under pressure, provided that this is not so high as to cause vapors
to condense. The vessel is evacuated and flushed several times
with the diluent, and then the evacuated vessel is connected to a
burette containing the appropriate amount of sample, which has a
liquid (e.g., mercury) closure. The mercury is prevented from
entering the vessel by a ball floating on its surface, which closes
against a sealing. Then the diluent gas is added to a set pressure
read from an accurate gauge.

A deficiency common to static methods is the scope for con-siderable error due to adsorption of trace components; this loss can attain 50% [5]. This makes it better to use dynamic methods when trace components are involved; sometimes these are the only ones possible. In these methods there is always dynamic equilibrium between the compounds and the adsorbing surfaces. The equilibration time is governed by the material used in the apparatus. Glass is the commonest material, which should be treated with ammonia or alkali to minimize the uptake [6]. Mix-tures involving certain corrosive gases at levels down to $10^{-7}\%$ equilibrate with glass and PTFE equipment in a few minutes, whereas hours or days may be needed for corrosion-resistant equipment [7].

A mixture with relatively high concentrations can be prepared by saturating a gas with the vapors of substances dissolved in a liquid [8] (the vapor pressure over the solution is calculated from Henry's law), or a gas may be saturated while bubbling at a con-stant rate through a temperature-cooled liquid [9] (Fig. 56), the temperature serving to adjust the concentration.

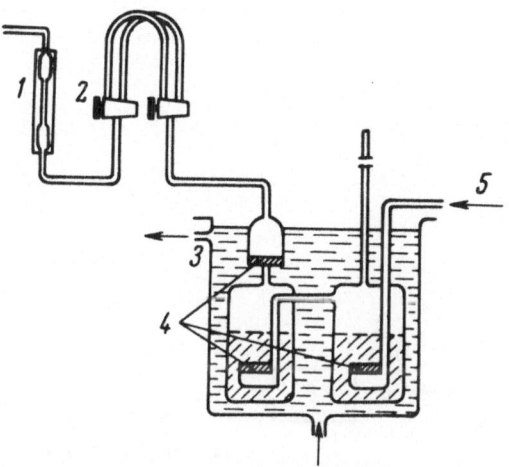

Fig. 56. The dynamic method of preparing stan-dard mixtures [9]: 1) rotameter, 2) calibrated space for sampling, 3) thermostat, 4) filters, 5) carrier-gas inlet.

Flow systems have been used [5, 7, 10, 11, 28, 29] to prepare gas mixtures. The system of [10] has three parallel tubes joined by capillaries (length 25 m, diameter 0.127 mm). The first tube contains the undiluted sample, which flows at about 5 cm³/min into the second tube, where the carrier gas flows at 2 liters/min. The diluted material passes through the second capillary to the third tube, where the required concentration is produced. Mixtures containing about 10^{-7} vol.% have been made in this way. A major disadvantage is that the pressure must be controlled very precisely in order to obtain reproducible results. This deficiency is common to all flow methods but is overcome in a gas mixer [12], which maintains a constant ratio of flow rates for the sample gas and carrier automatically.

Dynamic systems with diffusion-type dispensing are the most common; they are simple and reliable, while providing good accuracy and reproducibility if certain precautions are taken.

Lovelock [13-15] described a method of exponential dilution, where an initial gas sample is diluted by a carrier-gas (flow rate u, Fig. 57) in a mixer of volume V, which has a magnetic stirrer to maintain the same concentration in all parts of the vessel. If the substance is not adsorbed by the vessel, the concentration after a time t is

$$c = c_0 \exp\left[-\frac{Vt}{u}\right], \tag{116}$$

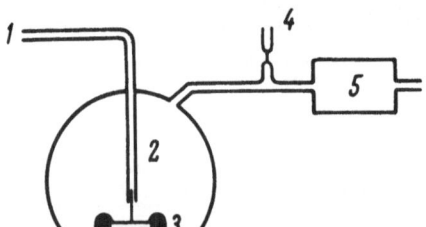

Fig. 57. Exponential dilution: 1) gas inlet, 2) mixer, 3) magnetic stirrer, 4) bypass, 5) detector.

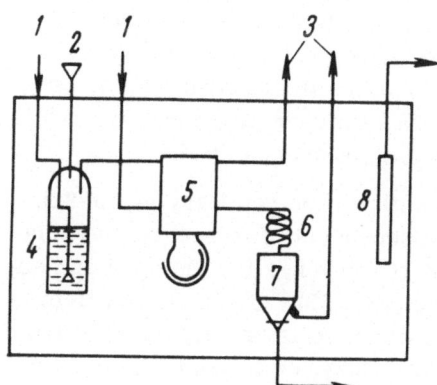

Fig. 58. Improved exponential-dilution technique [16]: 1) gas inlet, 2) sample inlet, 3) outlet to rotameters, 4) bubble monitor, 5) automatic dispensing valve, 6) column, 7) detector, 8) resistance thermometer.

where c_0 is the initial concentration. This has disadvantages similar to those of static methods, i.e., substantial error due to sorption.

An improved exponential-dilution method [16] largely eliminates the effects of adsorption and allows one to prepare very small concentrations of several substances simultaneously. A basic element is a glass bubble monitor, which contains the compound (or mixture) in a suitable nonvolatile liquid (Fig. 58). This receives pure argon, which then passes via an automatic dispensing valve and flow monitor to the atmosphere (or for a set time through the column to the detector). The entire system is placed in a thermostat.

Diffusion methods [17, 18] are also largely unaffected by adsorption. The vapor of the compound is in equilibrium with a liquid in which it is readily soluble, and it diffuses via a short capillary into a flow of the diluting gas; there are two vessels joined by a capillary, the lower one containing the solution. The concentration of each component is given by

$$c = c_0 DS/Lv, \tag{117}$$

where c_0 is the vapor concentration at the lower end of the capillary, D is the diffusion coefficient for the vapor in the gas, S is the capillary cross section, L is length, and v is gas flow rate. Convection due to temperature variations is minimized by an auxiliary

capillary, whose flow rate is 7-15 times that in the working capillary [18]. Standard mixtures with concentrations of 10^{-4} to $10^{-5}\%$
(ethane, propane, n-butane, and isobutane in nitrogen) can be prepared with a reproducibility of 5-15% (in operation for 7 hr), while
the error over a period of 3 days did not exceed 30-40%.

The liquid may be placed directly in the capillary [4, 19-22],
and then the results are largely independent of temperature variations [20]. The capillary is about 10 cm long, and the open end enters a mixer that receives the gas (Fig. 59). The entire system is
placed in a thermostat. The diffusion rate is readily deduced from
the speed of the meniscus in the capillary, or it can be calculated
from

$$r = \frac{2.30\,DMPS}{RTL}\log\frac{P}{P-p}, \tag{118}$$

where r is the diffusion rate (g/sec), D is the coefficient of molecular diffusion in the gas (cm^2/sec), M is the molecular weight of

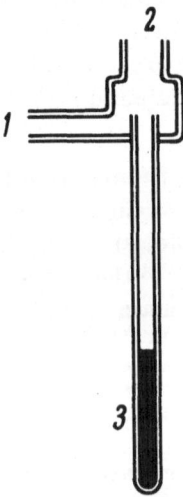

Fig. 59. Diffusion method of preparing mixtures:
1) gas inlet, 2) outlet to detector, 3) capillary
containing substance.

the vapor, P is the total pressure in the diffusion system (atm), p is the partial pressure (atm) of the vapor at temperature T, R is the gas constant (liter-atm/mole-deg K), and L is diffusion length (cm).

This method has been used [4] to produce mixtures with benzene and toluene concentrations of $0.01-0.5 \cdot 10^{-4}\%$.

The main disadvantage of the method is that the diffusion flux is not constant because L increases as the liquid evaporates. This disadvantage has been overcome [22] by using a capillary with two U bends, one (vertical) acting as the working part and the other (horizontal) acting as the measuring part. Here L is constant, and hence also is the diffusion flux, because the working part is U-shaped and the measuring part is horizontal.

An interesting method employs a layer of sorbent between the gas flow and the vapor source, which produces a constant concentration in the mixture and reduces the sensitivity to the hydraulic conditions in the mixing zone. The concentration is adjusted either via the gas flow rate or via the parameters that control the desorption (layer thickness, temperature, etc.).

Another diffusion technique [23] is of interest. The steady rate of diffusion of ethylene and propylene through the wall of a PTFE tube is only slightly dependent on gas composition and pressure. Tube permeabilities have been measured [23] for sulfur dioxide, hydrogen sulfide, nitrogen peroxide, propylene, butylene, and other hydrocarbons. Various applications of the method have been considered.

Sometimes very low concentrations in a flow or closed volume can be produced by electrolysis [9, 24]. Major advantages are the absence of working parts and the ease of control of the production rate (via the current). Electrolytic systems have been described [25] for generating oxygen, hydrogen, deuterium, nitrogen, chlorine, carbon dioxide, and nitric oxide.

Sometimes it is useful to employ a comparative method of calibration, with a flow divider and a less sensitive detector having known characteristics. The method was originally described [26] for calibrating an electron-capture detector, but its uses are more general. Figure 60 shows the system. The sample is injected into the column with a calibrated microsyringe, while the outlet has a

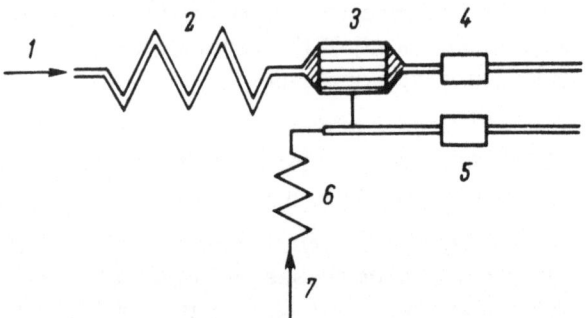

Fig. 60. Device for calibrating detectors: 1) gas inlet, 2) column, 3) isokinetic flow divider, 4) standard detector, 5) detector under calibration, 6) capillary, 7) gas entry for 5.

divider consisting of 20 capillary tubes. A small (known) part of the flow from the divider goes to the electron-capture detector, while the rest goes to the cross-section detector. The balance between the two flows can be adjusted within wide limits. A similar system can be used to calibrate a flame-ionization detector with a katharometer as standard detector.

We consider that it is often useful to prepare calibration mixtures in a standard chromatograph having a calibrated detector (e.g., a katharometer or flame-ionization detector), which may be connected in series with a vessel in which the standard mixture is prepared. This method can be applied to liquid and gaseous standard mixtures.

Kollerov [27] has discussed the theoretical principles of most of the above methods, as well as standard preparation at high concentrations.

LITERATURE CITED

1.   L. Batt and F. R. Cruickshank, J. Chromat., 21:296 (1966).
2.   M. L. Vlodarets, V. P. Bobrova, A. M. Rodionov, A. M. Popov, and M. L. Ternovskaya, Gas Chromatography, Proceedings of the Second All-Union Conference [in Russian], Moscow, Nauka (1964), p. 261.

3. J. Janák and K. Tesařík, Chem. Listy, 48:1051 (1954).
4. A. P. Altshuller and C. A. Clemons, Anal. Chem., 34:466 (1962).
5. B. E. Saltzman, Anal. Chem., 33:1100 (1961).
6. D. K. Kollerov, Physicochemical Measurements [in Russian], Trudy VNIIM, No. 68, Moscow—Leningrad, Standartgiz (1963).
7. P. D. Snelle, Instruments Speed Automation, No. 4, 128 (1957).
8. M. G. Burnett and P. A. T. Swoboda, Anal. Chem., 34:1162 (1962).
9. R. Aubeau, L. Champeix, and J. Reiss, J. Chromat., 16:7 (1964).
10. F. H. Huyten, J. W. Bijnders, and W. V. Beersum, Fourth International Gas-Chromatography Symposium (Preprints), 1962, Hamburg, edited by M. van Swaay, London, Butterworths (1962), p. 18.
11. G. S. Turner and W. M. Crum, Chem. Process (Engl.), 10:44 (1964).
12. Yu. P. Zolkin, Authors' certificate 124,108 (1960); Byull. Izobr., No. 22, 67 (1959).
13. J. E. Lovelock, Gas Chromatography, edited by R. P. W. Scott, London, Butterworths (1960), p. 16.
14. J. E. Lovelock, Anal. Chem., 33:162 (1961).
15. J. E. Lovelock, J. Chromat., 1:25 (1958).
16. V. Svojanovský, M. Krejči, K. Tesařík, and J. Janák, Chromat. Reviews, 8:90 (1966).
17. J. M. McKalvey and H. E. Hoelscher, Anal. Chem., 29:123 (1957).
18. A. A. Zhukhovitskii, N. M. Turkel'táub, and A. F. Shlyakhov, Neftekhimiya, 4:645 (1964).
19. J. M. N. Fortuin, Anal. Chim. Acta, 15:521 (1956).
20. D. H. Desty, C. J. Geach, and A. Goldup, Gas Chromatography, edited by R. P. W. Scott, London, Butterworths (1960), p. 46.
21. A. P. Altshuller and I. R. Cohen, Anal. Chem., 32:802 (1960).
22. O. K. Apukhtin, Author's certificate 166,163 (1963); Byull. Izobr., No. 21, 49 (1964).
23. A. E. Okeeffe and G. C. Ortman, Anal. Chem., 38:760 (1966).
24. M. Přibyl, Z. Anal. Chem., 217:7 (1966).
25. P. Hersch, C. Sambucetti, and R. Deuringer, Analysis Instrumentation — 1963, edited by L. Fowler, R. D. Eanes, and T. J. Kehoe, New York, Plenum (1963).
26. J. E. Lovelock, Anal. Chem., 35:474 (1963).
27. D. K. Kollerov, Metrological Principles of Analytical Gas Measurements [in Russian], Moscow, Committee on Standards, Measures, and Measuring Instruments, Council of Ministers of the USSR (1967).
28. J. P. Yquel, French patent, class Coln, No. 1,447,413.
29. L. Angely, E. Levart, G. Guiochon, and G. Peslerbe, Anal. Chem., 41:1446 (1969).